A GUIDE TO CYBERSECURITY FOR WATER AND WASTEWATER UTILITIES

By Steve Mustard, PE, CAP, GICSP

2024

Water Environment Federation
601 Wythe Street
Alexandria, VA 22314–1994 USA
https://www.wef.org

ISBN 978-1-57278-471-0

About WEF

The Water Environment Federation (WEF) is a not-for-profit technical and educational organization of 30,000 individual members and 75 affiliated Member Associations representing water quality professionals around the world. Since 1928, WEF and its members have protected public health and the environment. As a global water sector leader, our mission is to connect water professionals; enrich the expertise of water professionals; increase the awareness of the impact and value of water; and provide a platform for water sector innovation. To learn more, visit www.wef.org.

Dedication

To the unsung heroes who work every day to provide drinking water and sanitary services safely, securely, and reliably throughout the United States of America. May your dedication be recognized, and your contributions celebrated.

To the 2 billion people around the world without access to clean and safe drinking water, and the 4 billion people who lack adequate sanitation services. Your daily struggles are a testament to strength and hope. We stand united in acknowledgment and empathy.

Contents

List of Figures

List of Tables

Acknowledgments

In writing a book focused on water and wastewater utilities, I would like to acknowledge all the people who I have worked with in the global water sector.

To Neil Tubman, Mark Davison, June Irving, Richard Mead-Briggs, Tom Clarke, Rob Houghton, Peter Fruin, and Paul Harding for starting MetaSphere with me, and to Phil Tomlinson for making it bigger and better.

To Alison Hardy and Laurie Reynolds for the Institution of Engineering and Technology Control and Automation Technical and Professional Network, especially the fun times at the annual water event.

To Pete Easey, Malcolm Tyler, Mark Stansfield, Steve Severn, Steve Beadle, and Stuart Combellack for starting the UK Water Industry Telemetry Standards (WITS) project.

To Simon Harrison, Brian Lumb, Tony Miller, Leigh Taylor, John Parsons, Mike Wolff, and Paul Evans in the United Kingdom; Lar Spain, Willie Dempsey, Brian Martin, and Ciaran Martin in Ireland; and Neil Partis, Petra Crowther, Mike Wassell, Graeme Jones, Kev Star, and David Hope in Australia.

To Mark Lambert, Chris Davidson, John Guidone, BJ Hutchison, John Frashure, Erika Hoffman, Sandra Roberts, David Boyle, Tony Andrews, and John Howell in the United States.

To the staff of the International Society of Automation for working together with the member community to create a better world through automation. Special thanks to Claire Fallon, Morgan Foor, Liz Neiman, and, of course, Andrea Holovach.

To Mark Taylor, for believing in someone you met in a brewery and deploying automation technology to make your water system work better.

To Jim O'Dowd, for your review, your support, and giving me the opportunity to fish around in sewers for remote terminal unit components.

To Steve Huffman, Steve Pflantz, and Leo Staples for the review and continued efforts to find a better way to help the U.S. water and wastewater sector.

To Bob Radvanovksy for the review and long Saturday afternoon chats about critical infrastructure protection.

To Matt Holmes, Charles Stephens, Michael Preston, Mark Pepper, Kathy Weinsaft, and everyone at the National Rural Water Association and State Rural Water associations around the country.

To Barry Liner and Lisa McFadden for getting me started on the book writing journey.

To Brian Hill for the idea, support, and hard work converting my writing into a readable book.

About the Author

Steve Mustard is an independent automation consultant with extensive technical and management experience across multiple sectors.

Backed by more than 35 years of engineering experience, Mustard specializes in the development and management of real-time embedded equipment and automation systems. He serves as president of National Automation, Inc. Mustard is a recognized authority on industrial cybersecurity, having developed and delivered cybersecurity management systems, procedures, training, and guidance to multiple critical infrastructure organizations.

Mustard is a licensed professional engineer in Texas and Kansas, a UK-registered chartered engineer, a European-registered Eur Ing, an International Society of Automation (ISA) Certified Automation Professional®, a certified Global Industrial Cybersecurity Professional, and a Certified Mission Critical Professional. He is a fellow in the Institution of Engineering and Technology, a senior member of ISA, a member of the Safety and Security Committee of the Water Environment Federation (WEF), and a member of the American Water Works Association.

Mustard writes and presents on a wide array of technical topics and is the author of *Industrial Cybersecurity: Case Studies and Best Practices* and *Mission Critical Operations Primer*, both published by ISA. He has also contributed to other technical books, including WEF's *Design of Water Resource Recovery Facilities* (Manual of Practice No.8, 6th edition) and *The Digital Twin* published by Springer.

Chapter 1

Introduction

1.0 CYBERSECURITY IN THE WATER AND WASTEWATER SECTOR

This nation faces ongoing cybersecurity attacks that threaten its critical infrastructure. This is true in the water and wastewater sector, and perhaps especially so in rural water systems. Despite providing a life-sustaining resource for the community, these systems are afforded only limited resources to mitigate cybersecurity risks.

From the early 2000s, when Vitek Boden used a stolen laptop and a radio to wreak havoc at a water resource recovery facility in Queensland, Australia, through to recent incidents, such as the February 2021 incident at Oldsmar, Florida, wherein the level of sodium hydroxide was changed to toxic levels, either by mistake or intentionally, cybersecurity risks for water and wastewater operators are a constant threat to the critical national infrastructure.[1]

Government and private bodies alike have sought to address concerns, but to date, the solutions have served only to increase the administrative burden. The methodologies used to date are too complex for the rural water sector. Nothing has been successful in truly yielding any substantial decrease in cybersecurity risk. This means that the sector continues to present a threat to the security and well-being of the nation.

---

[1]The incident in Oldsmar, Florida, was originally reported as an intentional cybersecurity attack. It was subsequently reported that it was an operator error. However, both are forms of cybersecurity incidents that require management. This is discussed further in Chapter 4, Cybersecurity Barriers.

The effect of a ransomware incident in a rural water system could wipe out an entire annual budget. More importantly, high-profile incidents such as that in Oldsmar, Florida, show that there is the potential for widespread harm to the public and significant damage to the water and wastewater infrastructure such as that seen in Flint, Michigan. Although the damage in Flint was not a consequence of a cybersecurity incident, a similar result would have been possible by changing the chemical dosing in the water supply by malicious or accidental means.

The United States has more than 145,000 active public water systems providing water and wastewater services. Of these, 97% are considered small systems, meaning they serve 10,000 or fewer people, according to the Safe Drinking Water Act.[2] These smaller systems have limited resources to manage risks to their operations. As noted in the Water Environment Federation's (WEF's) Manual of Practice No. 33, *Information Technology for Water and Wastewater Utilities*, "a large portion of the populace do not have the knowledge or skills to recognize or guard against even the most rudimentary cyberattack."[3]

America's Water Infrastructure Act was signed into law in October 2018. This requires public water systems to conduct and develop vulnerability assessments and emergency response plans considering resiliency, cybersecurity, and physical security.[4]

According to information about these incidents obtained from the Cybersecurity & Infrastructure Security Agency[5] and the SCIDMARK database maintained by the Infracritical Organization,[6] between 2019 and 2021 alone there were at least seven reported incidents in U.S. water and wastewater utilities (WWUs). Four involved the use of ransomware and, in one of those cases, a wastewater treatment facility required manual operation until the computer system was restored. In the other three cases, employees (one of whom was a former employee who still had access to systems

---

[2]U.S. Environmental Protection Agency. (n.d.). *Safe Drinking Water Act (SDWA)*. Retrieved January 21, 2024, from https://www.epa.gov/sdwa

[3]Water Environment Federation. (2022). *Information technology for water and wastewater utilities* (2nd ed.; WEF Manual of Practice No. 33), p. 179.

[4]U.S. Environmental Protection Agency. (n.d.). *America's Water Infrastructure Act of 2018*. Retrieved January 21, 2024, from https://www.epa.gov/ground-water-and-drinking-water/americas-water-infrastructure-act-2018-awia

[5]Cybersecurity & Infrastructure Security Agency. (2021, October 25). *Ongoing cyber threats to U.S. Water and wastewater systems*. Retrieved January 21, 2024, from https://www.cisa.gov/news-events/cybersecurity-advisories/aa21-287a

[6]Infracritical. (n.d.). *Systems and cyber impact database markup*. Retrieved January 21, 2024, from http://search.infracritical.com/

despite having left the organization) accidentally or intentionally disrupted operations using their system access.

In March 2023, in response to the continued threat, the U.S. Environmental Protection Agency (U.S. EPA) issued a new memorandum ordering all public water systems to meet a series of basic cybersecurity requirements while also making cybersecurity audits a part of regular scheduled sanitary surveys.[7] The memorandum was withdrawn in October 2023, following a lawsuit involving the states of Missouri, Arkansas, and Iowa, as well as the National Rural Water Association and the American Water Works Association.[8]

Also in 2023, data from a U.S. EPA assessment were released. The assessment involved 249 public water systems that volunteered to be reviewed in three stages to assess their cybersecurity readiness over a 12-month period.[9] The data are self-reported and unverified and represent a tiny portion of the 145,000 public water systems in the United States. Although the survey indicates some positive movement toward better cybersecurity readiness, it reveals that there is still a very long way to go until the U.S. water and wastewater sector is adequately managing its cybersecurity risk.

The survey results are not surprising. Cybersecurity guidance is often complex and can be difficult to conduct in the field. There are a multitude of standards, guides, and frameworks, some that are even specific to the water and wastewater sector. Multiple government agencies, including the U.S. EPA, have an oversight role in cybersecurity. Experts will generally provide lists of more than 20 technical related points to be investigated, many that require the skills of a specialist to execute. Although there are many service providers in this area, they are driven by business interests, focusing services on technology, not people and process. For real change to take place, solutions must address people, process, and *then* technology. More importantly, WWUs need practical help to understand how to address this challenge.

This book recognizes the challenge that WWUs face. Successfully managing cybersecurity requires knowledge, skills, and resources targeted to reduce risk as efficiently as possible. Many WWUs have limited resources

---

[7]U.S. Environmental Protection Agency. (2023, March 3). *EPA takes action to improve cybersecurity resilience for public water systems.* Retrieved January 21, 2024, https://www.epa.gov/newsreleases/epa-takes-action-improve-cybersecurity-resilience-public-water-systems

[8]U.S. Environmental Protection Agency. (n.d.). *Cybersecurity in sanitary surveys.* Retrieved January 21, 2024, from https://www.epa.gov/waterresilience/cybersecurity-sanitary-surveys

[9]Geller, E. (2023, July 25). Exclusive: America is struggling to safeguard water supply from hackers, new EPA data shows. *The Messenger.* Retrieved September 10, 2023, from https://themessenger.com/tech/exclusive-america-is-struggling-to-safeguard-water-supply-from-hackers-new-epa-data-shows

and expertise. Despite the many forms of guidance, it remains difficult to know where to begin, and where to focus efforts.

2.0 ABOUT THIS BOOK

This book is intended to help those responsible for water or wastewater operations navigate the complex world of cybersecurity. The book assumes basic knowledge of technology and focuses on key issues that readers should consider.

The term *water and wastewater utility* is used throughout this book to refer to any entity that meets one of the following criteria:

- Community water system: A public water system that supplies water to the same population year-round.
- Non-transient non-community water system: A public water system that regularly supplies water to at least 25 of the same people at least 6 months per year. Examples include schools, factories, and hospitals that have their own water systems.
- Transient non-community water system: A public water system that provides water in a place such as a gas station or campground where people do not remain for long periods of time.

These systems may or may not provide wastewater treatment services.

This book is intended for use with WWUs of any size:

- Very small water systems serving 25 to 500 people.
- Small water systems serving 501 to 3,300 people.
- Medium water systems serving 3,301 to 10,000 people.
- Large water systems serving 10,001 to 100,000 people.
- Very large water systems serving more than 100,000 people.

Although larger systems may have more resources and capabilities than their smaller counterparts, the methodology described in this book applies equally well to any system.

The book begins with an introduction to the cybersecurity risk to WWUs. Whereas the number of publicly reported issues may seem low, it is crucial to understand the potential consequences of a cybersecurity incident in a WWU. There are many reasons why issues go unreported, including lack of awareness of incident cause, concern over repeat attacks, and in the case of WWUs, concern over bond devaluation and other reputational effects.

Having described the risk in Chapter 2, Chapter 3 then discusses one of the biggest challenges that WWUs face: knowing where to focus their limited resources to achieve the best possible result. Chapter 3 discusses how to do this, through a clearly defined risk assessment process.

Chapter 4 provides more details of the cybersecurity controls that should be in place in a WWU and gives guidance on the issues around these controls.

One of the most important things that a WWU can do to manage cybersecurity risk is to raise awareness within their team. Everyone can play a part in reducing cybersecurity risk. Chapter 5 gives guidance on how to develop a culture of good cybersecurity management.

Water and wastewater utilities do not operate in isolation. Third parties, such as engineering companies, system integrators, product vendors, and others interact on a regular basis. No matter how well the WWU manages their cybersecurity risk, if third parties do not manage their risk, the WWU remains vulnerable. Chapter 6 gives guidance on how to work with third parties to ensure they are managing their cybersecurity risk.

Cybersecurity risk management is a continuous process requiring constant vigilance and focus. Chapter 7 discusses how WWUs should maintain cybersecurity management.

Finally, Chapter 8 provides a list of recommended further reading and resources, including various assessment tools that can be used in the risk assessment process.

Chapter 2

The Cybersecurity Risk to Water and Wastewater Utilities

1.0 INTRODUCTION

Water and wastewater systems are one of the 16 sectors in the United States' critical infrastructure. The Department of Homeland Security (DHS) works to improve the security of the United States. The department's work includes customs, border, and immigration enforcement; emergency response to natural and human-made disasters; antiterrorism work; and cybersecurity. The Department of Homeland Security notes that: "Safe drinking water is a prerequisite for protecting public health and all human activity. Properly treated wastewater is vital for preventing disease and protecting the environment.

Thus, ensuring the supply of drinking water and wastewater treatment and service is essential to modern life and the Nation's economy."[1]

The Cybersecurity & Infrastructure Security Agency (CISA) is a component of DHS responsible for cybersecurity and infrastructure protection.The Cybersecurity & Infrastructure Security Agency recognizes water—along with energy, communications, and transportation—as the four infrastructure sectors that are "critical to the operations of all other sectors, and as such are fundamental to the delivery of the basic societal functions communities seek to provide."[2]

To reinforce this further, critical infrastructure experts Robert Radvanovsky and Allan McDougal identify water and wastewater as a sector that is "critical to a life-sustaining environment" and highlight examples in Cape Town, South Africa, and California, United States, where a lack of water challenges the order of society.[3]

Water and wastewater entities are not immune to a variety of cyber-induced incidents, including

- loss or theft of confidential customer data or regulatory reporting data;
- deliberate or accidental loss of service, resulting in loss of revenue and possible damage to reputation; and
- deliberate or accidental inappropriate use of equipment, resulting in harm to employees or the public, equipment, or the environment.

Consequently, critical services, such as firefighting and healthcare, as well as other sectors such as energy, food and agriculture, and transportation systems, can suffer significant negative effects from compromised water and wastewater systems.

2.0 REPORTED INCIDENTS IN THE WATER AND WASTEWATER SECTOR

Table 2.1 provides a list of known or reported incidents related to the U.S. water and wastewater sector. Note that for a variety of reasons, not

[1]Cybersecurity & Infrastructure Security Agency. (n.d.). *Critical infrastructure sectors.* Retrieved January 21, 2024, from https://www.cisa.gov/topics/critical-infrastructure-security-and-resilience/critical-infrastructure-sectors

[2]Cybersecurity & Infrastructure Security Agency. (n.d.). *Critical infrastructure systems.* Retrieved January 21, 2024, from https://www.cisa.gov/topics/critical-infrastructure-security-and-resilience/resilience-services/infrastructure-dependency-primer/learn/critical-infrastructure-systems.

[3]Radvanovsky, R., & McDougall, A. (2023). *Critical infrastructure: Homeland security and emergency preparedness* (5th ed.; pp. 165–166). CRC Press.

all incidents are reported. In some cases, incidents were not attributed to a cybersecurity cause. In other cases, incidents were not reported due to confidentiality concerns. As a result, this list is likely incomplete.

The information about these incidents is obtained from CISA,[4] the Repository of Industrial Security Incidents,[5] and the Systems and Cyber Impact Database Markup database maintained by the Infracritical Organization.[6]

3.0 CYBERSECURITY INCIDENTS IN WATER AND WASTEWATER UTILITIES

3.1 Contributing Factors

Water and wastewater utilities (WWUs) are constantly at risk of being affected by a cybersecurity incident, even if they are unaware of this fact. This risk applies to every WWU that has any programmable electronic equipment, such as an office personal computer, a programmable logic controller (PLC), a supervisory control and data acquisition (SCADA) system, a smart meter system, or even a billing system. Other devices or systems may be on the same network as water system equipment. For example, heating, ventilation, and air conditioning systems, closed-circuit television cameras, and Internet of Things devices such as those in security systems and smart lighting.

Although the risk is reduced if the programmable electronic equipment is not connected to the Internet, there is never zero risk.

Cybersecurity risk comprises three factors:

- threat,
- vulnerability, and
- exploit.

For a cybersecurity incident to occur, there first needs to be a *threat*. This can be a person, organization, circumstance, or event. There must also be a *vulnerability*. This could be a flaw in software or hardware, a configuration error, or a failure in a process or procedure. Finally, there must be a way to *exploit*, or use, the vulnerability, either intentionally or unintentionally.

[4]Cybersecurity & Infrastructure Security Agency. (n.d.). *Latest CRSB report*. Retrieved January 21, 2024, from https://www.cisa.gov

[5]Repository of Industrial Security Incidents. (n.d.). *RISI online incidents database*. Retrieved January 21, 2024, from https://www.risidata.com/

[6]Infracritical. (n.d.). *Systems and cyber impact database markup*. Retrieved January 21, 2024, from http://search.infracritical.com/

TABLE 2.1 List of Reported Incidents in the Water and Wastewater Sector

Year	Location	Description
1994	Arizona, USA	A hacker gained unauthorized access to a system monitoring water levels in canals and deleted data, resulting in a loss of $40,000 and a loss of service.[7]
2000	Australia	A disgruntled former contractor used stolen equipment and remotely accessed a wastewater control system multiple times over a period of months and released more than 0.7 ML (200 000 gal) of raw wastewater into the local environment, causing extensive damage.[8]
2006	Pennsylvania, USA	A WRRF's systems were accessed and infected with malware using an employee's laptop, which was compromised earlier while accessing the Internet.[9]
2011	Texas, USA	An ethical hacker remotely gained unauthorized access to a water treatment control system and published screenshots online to prove how easily it can occur.[10]
2011	Arizona, USA	A disgruntled former employee shut down critical systems responsible for wastewater control.[11]
2013	New York, USA	Iranian hackers gained access to a sluice gate control system as part of a large-scale breach of U.S. organizations. The control system was offline at the time.[12]
2019	Kansas, USA	A disgruntled former employee attempted to remotely tamper with a water treatment facility. The employee pleaded guilty and was sentenced in 2021.[13]

(continued)

[7]Repository of Industrial Security Incidents. (n.d.). *Salt River Project hack*. Retrieved January 21, 2024, from https://www.risidata.com/Database/Detail/salt-river-project-hack

[8]Repository of Industrial Security Incidents. (n.d.). *Maroochy Shire sewage spill*. Retrieved January 21, 2024, from https://www.risidata.com/Database/Detail/maroochy-shire-sewage-spill

[9]Repository of Industrial Security Incidents. (n.d.). *Pennsylvania water company hack*. Retrieved January 21, 2024, from https://www.risidata.com/Database/Detail/pennsylvania_water_company_hack

[10]Repository of Industrial Security Incidents. (n.d.). *South Houston Water Treatment Plant hack*. Retrieved January 21, 2024, from https://www.risidata.com/Database/Detail/south_houston_water_treatment_plant_hack

[11]Infracritical. (2011). *Greenfield Wastewater Treatment Plant (2011)*. Retrieved January 21, 2024, from http://scid.infracritical.com/pages/archived/scid-00006/

[12]Cybersecurity & Infrastructure Security Agency. (2020, October 24). *Potential for Iranian cyber response to U.S. military strike in Baghdad*. Retrieved January 21, 2024, from https://www.cisa.gov/news-events/cybersecurity-advisories/aa20-006a

[13]Infracritical. (n.d.). *Ellsworth County Rural Water District No. 1 (2019)*. Retrieved January 21, 2024, from http://scid.infracritical.com/pages/archived/scid-00119/ and Cybersecurity & Infrastructure Security Agency. (2021, October 25). *Ongoing cyber threats to U.S. Water and wastewater Systems*. Retrieved January 21, 2024, from https://www.cisa.gov/news-events/cybersecurity-advisories/aa21-287a

TABLE 2.1 List of Reported Incidents in the Water and Wastewater Sector (*Continued*)

Year	Location	Description
2020	New Jersey, USA	Personnel at a New Jersey-based water resource recovery facility discovered ransomware had compromised files within their system.[14]
2021	Florida, USA	An authorized user made a mistake setting the levels of a toxic chemical in a water treatment facility.[15]
2021	California, USA	Attackers used ransomware against a California-based water resource recovery facility. The ransomware variant had been in the system for about a month and was discovered when three SCADA servers displayed a ransomware message.[16]
2021	Maine, USA	Attackers used remote access to introduce ransomware onto a Maine-based water resource recovery facility's SCADA computer. The treatment system was run manually using local control and more frequent operator rounds until the SCADA computer was restored.[17]
2021	Nevada, USA	Attackers used ransomware against a Nevada-based water resource recovery facility. The ransomware affected the SCADA system and backup systems.[18]
2021	California, USA	A disgruntled former contract operator used his personal computer to remotely access a water treatment facility to intentionally cause the safety and protection systems to malfunction. The operator was indicted in 2023 and is awaiting trial at the time of writing.[19]

[14]Cybersecurity & Infrastructure Security Agency. (2025, October 25). *Ongoing cyber threats to U.S. Water and wastewater Systems*. Retrieved January 21, 2024, from https://www.cisa.gov/news-events/cybersecurity-advisories/aa21-287a

[15]Infracritical. (n.d.). *City of Oldsmar (2021)*. Retrieved January 21, 2024, from http://scid.infracritical.com/pages/archived/scid-00113/

[16]Cybersecurity & Infrastructure Security Agency. (n.d.). *Ongoing cyber threats to U.S. water and wastewater systems*. Retrieved January 21, 2024, from https://www.cisa.gov/news-events/cybersecurity-advisories/aa21-287a

[17]ibid.

[18]ibid.

[19]Cybersecurity & Infrastructure Security Agency. (n.d.). *CISA Tabletop Exercise Package—Water Systems*. Retrieved January 21, 2024, from https://www.cisa.gov/sites/default/files/publications/Water%20Systems%20CTEP_20210920_508_%20v0.0.docx

If a cybersecurity incident occurs there is a *consequence* for the WWU. The relationship between these factors is shown in Figure 2.1.

3.2 Consequences

Table 2.2 shows a list of potential consequences of a cybersecurity incident in a WWU.

Some of these consequences were observed in the water and wastewater sector incidents listed in Section 2.0. Some of these consequences occurred due to causes other than cybersecurity.

For example, Flint, Michigan's, decision to change water source and failure to apply corrosion inhibitors in the water treatment process resulted in the leaching of lead from pipes in the distribution network, exposing 100,000 residents to elevated lead levels.[20] The same result could theoretically occur because of a cybersecurity cause.

Some of these consequences have been seen in other sectors. For instance, the Stuxnet cybersecurity incident caused damage to centrifuges used in uranium enrichment,[21] and a steel mill in Germany suffered "massive damage" to its blast furnace because of a cybersecurity incident.[22]

3.3 Threats and Exploits

Cyber incidents can occur for many reasons:

- Amateur hackers, with access to many online tools and resources, can find systems connected to the Internet and interfere with their operation.
- Professional hackers, with more skills and resources, can target organizations with ransomware and other disruptive techniques for profit.

FIGURE 2.1 Relationship Between Threat, Vulnerability, Exploit, and Consequence

[20]Kennedy, M. (2016, April 20). *Lead-laced water in Flint: A step-by-step look at the makings of a crisis. NPR*. Retrieved August 29, 2023, from https://www.npr.org/sections/thetwo-way/2016/04/20/465545378/lead-laced-water-in-flint-a-step-by-step-look-at-the-makings-of-a-crisis

[21]Langer, R. (2013, November). *To kill a centrifuge: A technical analysis of what Stuxnet's creators tried to achieve*. The Langner Group. Retrieved August 29, 2023, from https://www.langner.com/wp-content/uploads/2017/03/to-kill-a-centrifuge.pdf

[22]Hack attack causes 'massive damage' at steel works. (2014, December 22). *BBC*. Retrieved August 29, 2023, from https://www.bbc.com/news/technology-30575104

TABLE 2.2 List of Potential Consequences of a WWU Cybersecurity Incident

Consequence	Example(s)
Loss of potable water supply or sanitary services affecting the local community	Water scarcity Effect on other services (e.g., firefighting) Costly temporary solution (e.g., transportation of water from other areas)
Compromise of potable water supply resulting in harm to customers	Poisoning or skin irritation of members of the public
Damage to infrastructure	Pipe corrosion due to incorrect treatment Pipe burst due to incorrect pressure or flow control
Damage to machinery or other process equipment	Cavitation of pump or burnout of motor due to operation at low flow
Harm to the environment	Release of untreated wastewater or toxic chemicals into the water course
Harm to employees	Exposure to toxic chemicals (e.g., chlorine gas) due to a lack of supervisory control
Loss of customer or employee personally identifiable information (PII)	Theft of Social Security numbers, names, addresses, or bank details from payment system
Loss of support services such as customer billing or employee payment	Manual methods required (e.g., collecting or writing checks)

- Activist hackers can disrupt the operations of organizations that they consider are not aligned with their political or social views.
- Disgruntled employees or contractors can use inside knowledge, or privileged access, to gain revenge.
- Nation-states or terrorists can target critical infrastructure organizations to create instability.

Cyber incidents need not be the result of deliberate malicious intent. Employees or contractors can also cause serious consequences, for instance by simply failing to maintain adequate backups.

3.3.1 Amateur Hackers

Often referred to as "script kiddies" (or "skid" or "skiddie"), these are unskilled hackers who use freely available automated tools written by others (usually professional hackers) to break into computer systems.

Amateur hackers can be motivated for a wide range of reasons. They tend to be indiscriminate and try to compromise any system on the Internet they can reach. They might lack detailed technical knowledge, but they may explore systems simply for fun, or for recognition.

There is a wide range of information freely available on the Internet that amateur hackers can use to attack WWUs.

Shodan is a search engine for Internet-connected devices.[23] It allows users to perform complex searches, for instance within geographic locations, for types of system or communications protocols. From the results, users potentially obtain useful information that can be used in an attack. Figure 2.2 shows the results of a search showing what claims to be part of

FIGURE 2.2 Shodan Allows Anyone to Search for Internet Connected Devices (https://shodan.io)

[23] Infracritical. (n.d.). *Shodan*. Retrieved January 21, 2024, from https://shodan.io

a municipal water system. Some of the details have been redacted on the screenshot.

Other resources on the Internet, such as the Mitre Common Vulnerabilities and Exploits (CVE) database[24] and the National Institute of Standards and Technology (NIST) National Vulnerability Database (NVD)[25]provide details of known product vulnerabilities, or flaws, that can be used to gain access to WWUs. Figure 2.3 shows one such example, describing a flaw in a particular brand of PLC.[26] This flaw means that users can send a command

Module Overview

Name: [redacted]Remote START/STOP Command
Module: auxiliary/admin/scad[redacted]d
Source code: modules/auxiliary/admin/scada/[redacted]rb
Disclosure date: 2012-04-05
Last modification time: 2020-10-02 17:38:06 +0000
Supported architecture(s): -
Supported platform(s): -
Target service / protocol: -
Target network port(s): 502
List of CVEs: -

The[redacted]PLCs use Modbus function code 90 (0x5a) to perform administrative commands without authentication. This module allows a remote user to change the state of the PLC between STOP and RUN, allowing an attacker to end process control by the PLC. This module is based on the original [redacted]b' Basecamp module from [redacted]

Module Ranking and Traits

<u>Module Ranking</u>:

- **normal**: The exploit is otherwise reliable, but depends on a specific version and can't (or doesn't) reliably autodetect. More information about ranking can be found here.

Basic Usage

```
msf > use auxiliary/admin/scada/[redacted]d
msf auxiliary[redacted]nd) > show targets
    ... a list of targets ...
msf auxiliary([redacted]) > set TARGET target-id
msf auxiliary([redacted]) > show options
    ... show and set options ...
msf auxiliary([redacted]) > exploit
```

FIGURE 2.3 Online Resources Describe Methods to Exploit Vulnerabilities (https://www.infosecmatter.com)

[24]Mitre. (n.d.). *Common vulnerabilities and exploits*. Retrieved January 21, 2024, from https://cve.mitre.org/

[25]National Institute of Standards and Technology. (n.d.). *National vulnerability database*. Retrieved January 21, 2024, from https://nvd.nist.gov/

[26]InfosecMatter. (n.d.). *Metasploit*. Retrieved January 21, 2024, from https://www.infosecmatter.com/metasploit-module-library/?mm=auxiliary/admin/scada/modicon_command

to remotely start or stop the PLC operation. Some of the details have been redacted on the screenshot.

Amateur hackers can combine the knowledge from these and other resources to mount attacks on WWUs with very little skill or knowledge.

Some of the other freely available tools and resources online include

- Nmap (network mapper)—This is used to discover what devices and applications are running on a network.
- Nessus—This tool scans networks looking for software that has not been updated, weak passwords, or other known vulnerabilities.
- Wireshark—This is used to collect and analyze network data. Collecting data can lead to the discovery of unprotected passwords, as well as details about what devices and applications are running.
- John the Ripper—This tool is a password security auditing and password recovery tool available for many operating systems. It can be used to hack, or crack, passwords.

There are hundreds of other tools available online for amateur hackers to use at will.

3.3.2 Professional Hackers

There are various types of professional hacker:

- Ethical hacker—Formerly known as "white hat" hacker, this is a professional whose objective is to identify vulnerabilities and have them fixed before they are exploited by malicious hackers. They may be individual researchers, cybersecurity company employees, or government representatives.
- Unethical hacker—Formerly known as "black hat" hacker, this group includes individuals and organized crime groups, as well as nation-states and others. For the purposes of this discussion, nation-states and others are covered separately.

There are other types of professional hacker (red hats, grey hats, green hats, and so on) that exist between ethical and unethical types, but for the purposes of this book, ethical and unethical will be used.

Hacking is a big business. Some organizations have call centers that make unsolicited calls pretending to be well-known companies. They trick victims into giving remote access to their computers and then deploy malicious software (or malware). This might give them access to sensitive information, such as banking details or passwords, or may lock the user out until they pay a fee (this type of malware is called *ransomware*).

Some attacks are fully automated with no human intervention. Software searches the Internet for unprotected devices and can perform brute force password attacks, or automatically deploy the appropriate malicious code.

Ransomware is such big business that services exist for others to use. Online services ("malware as a service") exist in which any users (including amateur hackers) can create their own ransomware and have it deployed automatically. A fee (in this case 25%) is deducted from all the money collected and the remainder is paid to the user in some form of electronic currency (e.g., Bitcoin).[27]

A WWU may not seem like an attractive target to a professional hacker. However, it is not necessary to be a target to be the victim of a cybersecurity incident. For example, in May 2017, a ransomware cryptoworm known as WannaCry infected an estimated 230,000 computers in 150 countries in a matter of hours. WannaCry locked users out of their computers until a $300 ransom was paid. Organizations such as the UK National Health Service, French vehicle manufacturer Renault, and global shipping company Maersk (through a vendor) were all affected, but none were specifically targeted.

3.3.3 Activist Hackers

Also known as "hacktivists," these are activist hackers that use hacking as a method to promote their own cause or political agenda. Organizations in sectors such as oil and gas can be particularly at risk from hacktivists, seeking to disrupt operations they consider to be harmful to the environment. Whereas WWUs are not so controversial, there can still be a risk from hacktivists, perhaps because of concerns over extraction of water or disposal of wastewater in sites of special scientific interest. Attacks on WWUs can also be used to voice larger political messages, such as concerns about nation-level actions against minorities.

Hacktivists may themselves be professional hackers with a particular agenda or may team up with professional hackers to achieve their goals. As already noted, the range of resources available online means that it is not even necessary to team up with professional hackers.

Terrorists can be considered a type of hacktivist group, but for purposes of this discussion, they have been included with nation-states.

---

[27]Paganini, P. (2015, September 1). *ORX Locker, the new darknet ransomware-as-a-service platform*. Security Affairs. Retrieved January 21, 2024, from https://securityaffairs.com/39753/cyber-crime/orx-locker-raas.html

3.3.4 Disgruntled Employees or Contractors

Most cybersecurity discussions focus around amateur or professional hackers, outsiders who look to gain access to organizations to do harm or make money. Insufficient attention is paid to disgruntled employees or former employees. Contractors can be included in this category, as they can have the same access and information as employees.

The Department of Homeland Security highlights the need to address the risk of what they call the *insider threat*. They define an insider threat as "the threat that an employee or a contractor will use his or her authorized access, wittingly or unwittingly, to do harm to the security of the United States."[28]

A 2015 study by Intermedia revealed that 94% of office workers engage in some form of unsafe online habits that could jeopardize their employer or their customers. The same report found that for information technology professionals:

- 32% have given out their login/password credentials to other employees (compared to 19% across all respondents),
- 28% said they have accessed systems belonging to previous employers after they left the job (compared to only 13% among all respondents), and
- 31% said they would take data from their company if it would positively benefit them—nearly 3 times the rate of general business professionals.[29]

Despite this study's shocking findings, in the ensuing 5 years from that study, the problem only became worse. Research conducted in 2020 from the Ponemon Institute showed that the frequency and cost of insider threats have increased. They identified that malicious insider threats made up 14% of all incidents, but these carried a per-incident cost of $756,000, with annual losses of $4.08 million from all those surveyed.[30] Disgruntled employees and contractors remain a major threat to WWUs, but one that can be managed.

[28]Department of Homeland Security. (n.d.). *Insider threat.* Retrieved September 3, 2023, from https://www.dhs.gov/science-and-technology/cybersecurity-insider-threat

[29]Intermedia. (2015, October 21). *New data on security habits pinpoints the riskiest business users.* Retrieved September 3, 2023, from https://www.intermedia.com/press-release/new-data-on-security-habits-pinpoints-the-riskiest-business-users

[30]Cost of insider threats rises 31%. (2020, January 30). *Infosecurity Magazine.* Retrieved September 3, 2023, from https://www.infosecurity-magazine.com/news/cost-of-insider-threats-rises-31/

3.3.5 Nation-States and Terrorists

Nation-state cyberattacks originate from a particular country to further that country's interests. They may be part of an official state-sanctioned mission, members of a professional hacker outfit that is aligned with the government, or contracted by a government for a specific operation. The operation may be intended to steal intellectual property or destabilize another nation by disrupting critical infrastructure such as power, water, or transportation.

The groups involved in nation-state attacks are commonly known as advanced persistent threats (APTs). These organizations have sophisticated levels of cybersecurity expertise and significant resources.

The previously mentioned Stuxnet incident is an example of a nation-state attack. Although still unconfirmed, it is believed that the United States and Israel collaborated on the program to develop the malware to disrupt operations in an Iranian uranium enrichment facility.

A Russian APT group known as "Sandworm" is believed to have been behind a cyberattack on the Ukrainian power grid in 2015 that resulted in power outages for 230,000 customers for up to 6 hours.[31]

There can be some overlap between nation-states and terrorist groups. For example, The Taliban, Al Qaeda, ISIS, and Boko Haram are classified as terrorist groups by the U.S. State Department, but are also closely aligned with certain nations and ideologies and have the resources to deploy significant attacks on other nations.

Nation-states and terrorists pose a significant threat to WWUs. The 2013 incident in New York involving Iranian hackers accessing a sluice gate control system that was included in the list of published incidents is one example. A 2020 incident in Israel,[32] in which a coordinated attack on the control and monitoring systems of water resource recovery facilities, pumping stations, and water treatment facilities was conducted, highlights the potential for massive disruption to critical infrastructure.

More recently, in December 2023, an Iranian group known as "Cyber-Av3ngers" actively targeted and compromised Israeli-made PLCs used in the water and wastewater systems sector. The municipal water authority of Aliquippa, in Pennsylvania, was one of the U.S. water systems that was

[31] Finkle, J. (2016, January 7). *U.S. firm blames Russian 'Sandworm' hackers for Ukraine outage*. Reuters. Retrieved September 3, 2023, from https://www.reuters.com/article/us-ukraine-cybersecurity-sandworm-idUSKBN0UM00N20160108

[32] Council on Foreign Relations. (2020, May). *Attack on Israeli water utilities*. Retrieved January 21, 2024, from https://www.cfr.org/cyber-operations/attack-israeli-water-utilities

affected.[33] The municipality released a photograph showing a message on the PLC's display posted by the attackers.[34]

A Chinese state-sponsored APT group called Volt Typhoon is using legitimate network administration tools to target end-of-life network equipment still in operation within the U.S. critical infrastructure, including water and wastewater systems providing services to military bases.[35]

3.3.6 Unintentional Acts

Unintentional acts cause cybersecurity incidents. The U.S. Cybersecurity & Infrastructure Security Agency defines two types of unintentional threat: negligence and accidental. An example of a negligent act is a failure to follow policies or procedures, such as not installing security updates. An example of an accidental act is clicking on a link in a phishing email.[36]

Like disgruntled employees and contractors, the cybersecurity risk from unintentional non-malicious acts is not always fully addressed by organizations. However, survey after survey has consistently shown that unintentional non-malicious acts are responsible for most cybersecurity incidents. One such survey, from Elevate Security, showed that 85% of security incidents originated from such acts. Worse still, they identified that these incidents arose from the actions of fewer than 10% of users.[37] These statistics are consistent with dozens of other reports from other organizations.

Given the preponderance of data, the greatest cybersecurity risk to WWUs is from unintentional acts. As with disgruntled employees and contractor threats, it can be managed.

3.4 Vulnerabilities

As already noted, vulnerabilities are flaws. They can be one of three types:

- technical—a flaw in software or hardware, a configuration error;

---

[33] "Cybersecurity & Infrastructure Security Agency. (2023, December 1). IRGC-affiliated cyber actors exploit PLCs in multiple sectors, including U.S. water and wastewater systems facilities. https://www.cisa.gov/news-events/cybersecurity-advisories/aa23-335a

[34] Levy, M. (2024, January 2). After Aliquippa hack, Congress wrestles with cybersecurity at water utilities amid federal warnings. https://www.wesa.fm/politics-government/2024-01-02/aliquippa-water-utility-hack

[35] WaterISAC. (2023, December 14). *People's Republic of China state-sponsored cyber actor Volt Typhoon (Updated December 14, 2023).* Retrieved January 21, 2024, from https://www.waterisac.org/portal/peoples-republic-china-state-sponsored-cyber-actor-volt-typhoon

[36] Cybersecurity & Infrastructure Security Agency. (n.d.). *Defining insider threats.* Retrieved September 3, 2023, from https://www.cisa.gov/topics/physical-security/insider-threat-mitigation/defining-insider-threats

[37] Elevate Security. (n.d.). *The Top 3 cybersecurity incidents caused by unintentional insider risk.* Retrieved September 3, 2023, from https://elevatesecurity.com/the-top-3-cybersecurity-incidents-caused-by-unintentional-insider-risk/

- process—a failure in a process or procedure; or
- people—a failure to meet personal responsibilities.

3.4.1 Technical Vulnerabilities

The most common technical vulnerabilities in a WWU are shown in Table 2.3.

TABLE 2.3 List of Common Technical Vulnerabilities in WWUs

Vulnerability	Description
Insecure Internet Access	There may be nothing in place to control traffic to and from WWU equipment. This could mean that equipment is visible on the Internet. It can also make it easier for attackers to gain access to equipment. It may be unclear what equipment is connected to the Internet. Although equipment may not have a direct connection, there may be another route that can be used to gain access.
Out-of-Date Software	There are flaws in operating system software (such as Windows) and application software (such as billing programs, SCADA software) that can be exploited by attackers. These are reported by vendors, governmental agencies such as CISA, and other organizations such as WaterISAC. Fixes for these flaws are made available and must be installed if the vulnerability is to be removed.
No Malicious Code Protection	Flaws are constantly being discovered in software. Even the best-maintained equipment may still be vulnerable to malicious code. Some malicious code exploits configuration issues that are not addressed by software updates. Antivirus software is available for personal computers, mobile devices, and tablets and should be deployed to reduce the possibility of infection by malicious code.
Uncontrolled Use of USB Devices	Devices with USB ports, such as removable drives, phones, and tablets pose a significant risk to WWUs. These devices have their own storage, and this storage can be used to hold malicious code. This malicious code can be installed intentionally or unintentionally. Plugging in these devices into WWU equipment to either transfer data or recharge can allow malicious code to be transferred.
Uncontrolled Use of Portable Computers	Like USB devices, portable computers may not be kept up to date with security fixes and anti-malware protection and could be used intentionally or unintentionally to install malicious code if connected to WWU networks. This problem can be exacerbated using contractors or other third parties who may not follow the water system policies.

(*continued*)

TABLE 2.3 List of Common Technical Vulnerabilities in WWUs (*Continued*)

Vulnerability	Description
Uncontrolled Remote Access	Remote access is a useful tool that allows WWU employees to remotely monitor operations at site. Various types of applications are available to provide remote access. Some are built into Microsoft operating systems, some are freely available on the Internet, and others are paid solutions. If it is easy for a WWU employee to remotely access equipment at a site, it is probably easy for an unauthorized person to do the same.
Poor Design	The equipment that WWUs use should be designed to reduce the possibility that a cybersecurity incident can cause a serious situation. For example, although the 2021 incident in Florida mentioned earlier in this chapter did not result in a toxic level of chemical in the water supply, there was no limitation on the SCADA system to prevent a toxic level being set (no high or low limits, no error checking, and no confirmation mechanism). Other design flaws include deployment of software on shared machines, insufficient redundancy, and a lack of consideration of failure scenarios.
Poor Physical Access Control	Physical security is a key part of cybersecurity. As noted earlier in this chapter, cybersecurity risks are not limited to Internet-based attackers. Attackers will look for the easiest way to compromise a WWU. If that is by breaking into a building and accessing the equipment directly, then that is what they will do. An example of poor physical access control is to have rudimentary padlocks on operational sites, with a common key shared by multiple users. Good physical access control also reduces the possibility of unintentional incidents by limiting who can access key equipment.
Poor Password Management	Passwords exist to identify and authorize users to access WWU equipment. If these passwords are not adequately managed, their benefit is reduced. Examples of poor password management are • lack of multi-factor authentication; • use of a common username/password shared by multiple users; • password reuse across multiple systems or accounts; • failure to change password from default; • storage of passwords in Word, Excel, or text files on a personal computer; and • manufacturer "backdoor" accounts for support purposes.

3.4.2 Process Vulnerabilities

Processes and procedures play a crucial role in the functioning of a WWU. They are essential for maintaining consistency, efficiency, quality, and safety. They also play a crucial role in risk management, compliance, and the overall effectiveness of the WWU.

For cybersecurity risk management, WWUs will require processes and procedures to

- manage joiners, movers, and leavers, and ensure they have the access the need to do their job and remove them when no longer required;
- maintain equipment condition, including keeping backups of electronic equipment software, configuration, and program files;
- make changes to equipment and processes;
- respond to incidents, including cybersecurity incidents;
- check configuration file(s) for tampering;
- maintain secondary or tertiary backup equipment due to a cyberattack (e.g., replace "contaminated" equipment from ransomware); and
- conduct periodic audits of maintenance performed on or access to control systems.

Common process vulnerabilities include

- failure to establish a process—without a defined process, there is no way of ensuring that mistakes are not being made;
- failure to document or communicate the process—without a documented procedure that everyone understands, there is no way to ensure that employees do the right thing; and
- lack of oversight on the process—with no oversight, there is no way of knowing if employees are following the process.

Examples of the effect of process failures are

- a disgruntled employee leaving the WWU but still having physical keys or electronic account information,
- changes made to systems creating new cybersecurity vulnerabilities, and
- failure to perform backups resulting in inability to restore a system infected by ransomware.

3.4.3 People Vulnerabilities

People have the biggest effect in a WWU's cybersecurity preparedness. Without awareness, training, and good oversight, employees and contractors are the most likely cause of cybersecurity incidents. As noted earlier in this chapter, most cybersecurity incidents occur through unintentional acts from authorized users.

The most common unintentional acts that can lead to cybersecurity incidents are

- failure to identify a phishing email and either clicking on a link or downloading an attachment;
- failure to follow good security practices, such as connecting work or personal devices to untrusted universal serial bus (USB) ports (e.g., airports, hotels), use of insecure wireless networks (e.g., hotels, coffee shops) without a virtual private network;
- failure to check removable media (such as a USB drive) for malware before inserting it into an operational system computer. A "sheep-dip" process can be used to test such devices in a controlled and isolated manner;
- failure to adequately protect user credentials, such as sharing them with others, writing them down, or not updating them when required (e.g., after a cybersecurity incident);
- failure to physically secure equipment, such as leaving cabinet or room doors open or unlocked, including while traveling while traveling (e.g., failing to secure work or personal equipment in vehicle or hotel room); and
- failure to act when someone else is performing an insecure act, including any of the above.

3.4.3.1 Phishing

Countless surveys show that phishing continues to be a major issue. For example, according to CISA:

- Eight out of 10 organizations had at least one individual who fell victim to a phishing attempt in tests.
- Eighty-four percent of individuals replied with sensitive information or interacted with an attachment or link in a phishing email within the first 10 minutes of receiving it.
- Only 13% of individuals reported phishing attempts.[38]

[38]Cybersecurity & Infrastructure Security Agency. (n.d.). *Phishing*. Retrieved November 7, 2023, from https://www.cisa.gov/sites/default/files/2023-02/phishing-infographic-508c.pdf

Examples of phishing emails are regularly used in social engineering training.[39] These emails have common features and often appear to be from legitimate companies, but the sender email is not from the real company domain, the company logo is fake, and there are spelling or grammar errors. Links in these emails will either download malicious code to the user's machine or will take the user to a site where they are prompted to enter personal information, which is then used to compromise other accounts belonging to the user.

There are so many scams including phishing attempts involving companies such as Costco that they devote an entire page of their website with examples.[40] Many other companies have the same issue and provide similar information to help users spot phishing emails and other scams.

3.4.3.2 Removable Media

Removable media such as USB drives can become infected in a variety of ways. In some cases, someone can copy files from a machine without active anti-malware protection and malware can be transferred in the process, without the user's knowledge. In some cases, malware can be deployed on removable media intentionally.

For example, in January 2022, the FBI issued a security alert about FIN7, a cybercrime gang. The FBI said that over the previous several months, FIN7 mailed malicious USB devices containing a programmable microcontroller to companies in the hopes that somebody would plug in the drives, infect systems with malware, and thus set them up for future ransomware attacks.[41] The attackers disguised the packages as either pandemic-related (from the Department of Health and Human Services) or as a package from Amazon.

If someone fell for the attack, the FBI said that the devices were programmed to act as a keyboard preloaded, which automatically executed keystrokes on the victim's computer.

3.4.3.3 Passwords

Protecting user credentials continues to be another major issue. Organizations often suffer from systematic failures such as the following:

- Weak Passwords—People create weak passwords, such as "password, 123456," and so on. Passwords like this can be easily guessed using

[39]Security Metrics. (n.d.). *Social engineering training: What your employees should know.* Retrieved November 7, 2023, from https://www.securitymetrics.com/blog/social-engineering-training-what-your-employees-should-know

[40]Costco. (n.d.). *Currently known scams.* Retrieved November 7, 2023, from https://customerservice.costco.com/app/answers/answer_view/a_id/9770/~/currently-known-scams

[41]WaterISAC. (n.d.). *FBI FLASH: FIN7 cyber actors target US businesses through USB keystroke injection attacks.* Retrieved November 7, 2023, from https://www.waterisac.org/portal/fbi-flash-fin7-cyber-actors-target-us-businesses-through-usb-keystroke-injection-attacks

various techniques, such as brute force (in which a program tries all possible combinations of passwords until the correct one is found). Dictionary attacks, a form of brute force, uses a dictionary of commonly used passwords.

- Default Passwords—Many devices come with default credentials and users often do not change these. In some cases, these credentials are published in user guides or websites, making it easy for attackers to use them to gain unauthorized access to devices.
- Password Reuse—People routinely reuse passwords for multiple accounts and websites. If one account is compromised, it can lead to unauthorized access to other accounts. For example, if a WWU employee's personal email is compromised, and they use the same password for the WWU's billing application, then an attacker can use the employee's personal details to gain unauthorized access to the billing application. Credential stuffing is a type of attack that uses stolen username–password combinations from one breach to attempt unauthorized access to other accounts. Reuse of passwords makes this type of attack more likely to succeed.
- Social Engineering—Attackers manipulate individuals into disclosing sensitive information through various forms of social engineering, such as impersonation or deception. Phishing is a specific form of social engineering where attackers use specially crafted emails to trick users into revealing their user credentials.
- Password Management—Users often write down passwords. Figure 2.4 shows an example of a username and password on a label on a network switch (actual credentials redacted). Other examples include posting the password for the office Wi-Fi on a noticeboard that is visible from outside the office. Sometimes passwords are saved in an unprotected file on a computer or stored in a book in the office.

FIGURE 2.4 Username and Password on Equipment Label

- Inadequate Password Policies—Many WWUs have no policies in place regarding passwords. As a result, they may not enforce many controls that could address the above vulnerabilities.

Figure 2.5 shows the home page of the website *Have I Been Pwned?*[42] The website provides the ability to search data from security breaches to see if one's credentials are included. The site does not store usernames and

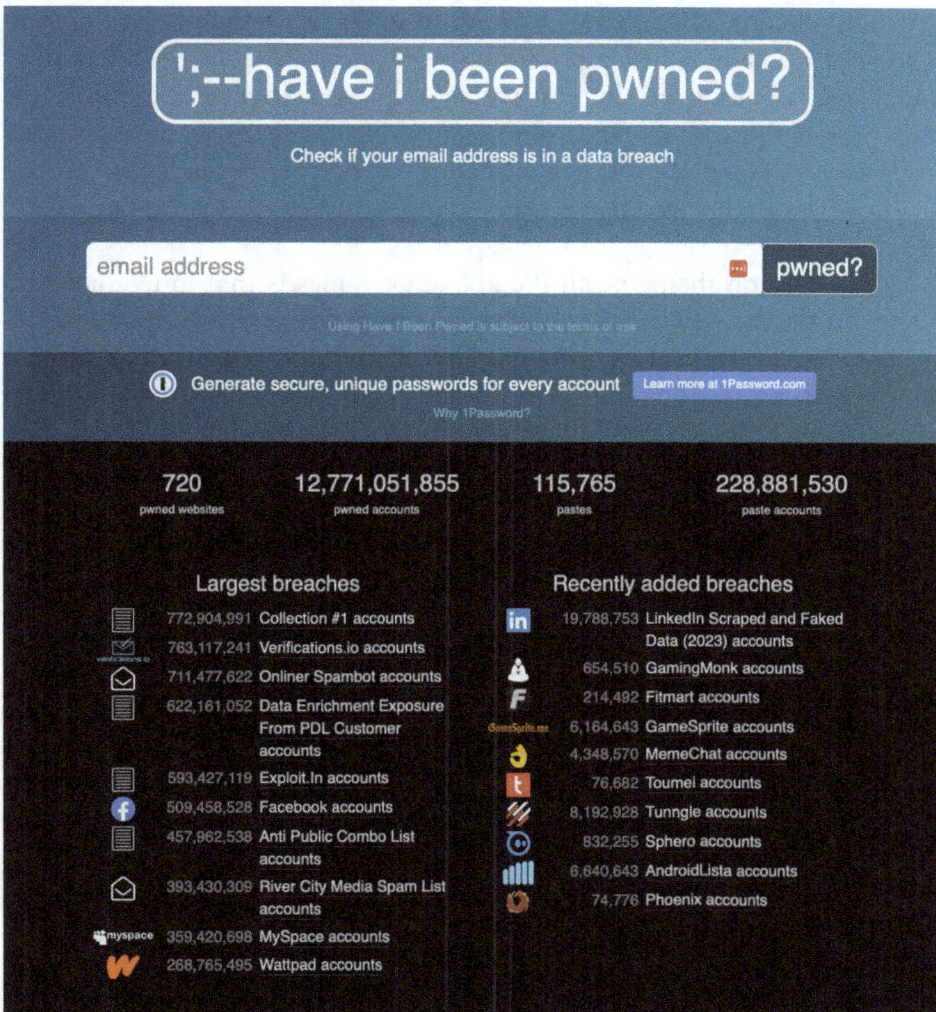

FIGURE 2.5 Have I Been Pwned? Website (https://haveibeenpwned.com/)

[42]"'. . . *have i been pwned?*" https://haveibeenpwned.com/, retrieved January 21, 2024. The word *pwned* has origins in video game culture and is a derivation of the word *owned*, used to imply that someone has been controlled or compromised.

passwords together so there is no potential security issue in the site itself. Users can also register their email and the site will notify them if it shows up in a future data breach.

3.4.3.4 Physical Security

Physical security is a key part of good cybersecurity management, but it is also subject to poor behaviors. Leaving lockable equipment cabinets unlocked, or leaving keys in the lock, is a common example. Another common example is shown in Figure 2.6. Here, a door is permanently kept unlocked so that people can get in and out of the secure room to take smoke breaks. Another poor physical security behavior is "tailgating." Tailgating is one of the simplest forms of a social engineering attack. An unauthorized party can get around security mechanisms that are assumed to be secure by simply following an authorized person through a barrier or door. The authorized person may not wish to challenge the individual or seem rude.

The common theme of all the above examples is that individuals can make a difference. Employees and contractors, who are aware of cybersecurity risks, are trained in processes and procedures, and who understand their role in protecting the WWU, are the single most effective contributor to secure operations. They will be more likely to spot suspicious emails; will scan removable media before use (and probably never use something that claims to come from Amazon); will be careful with their user credentials; and, most importantly, will more likely report any insecure behaviors from others, such as physical security failures.

FIGURE 2.6 An Example of Physical Security Control Failure

4.0 SUMMARY

This chapter has introduced the cybersecurity risk to WWUs. Water and wastewater systems are one of the 16 sectors in the United States' critical infrastructure, so the consequences of a cybersecurity incident can be significant.

Several cybersecurity incidents have been reported in water systems around the world. For a variety of reasons, not all incidents are reported, and it is important that WWUs do not dismiss the risk as negligible. Water and wastewater utilities are constantly at risk of being affected by a cybersecurity incident, even if they are unaware of this fact. This risk applies to every WWU that has any programmable electronic equipment. Although the risk is reduced if the programmable electronic equipment is not connected to the Internet, there is never zero risk.

For a cybersecurity incident to occur, there first needs to be a threat. This can be a person, organization, circumstance, or event. There must then be a vulnerability. This could be a flaw in software or hardware, a configuration error, or a failure in a process or procedure. Finally, there must be a way to exploit, or use, the vulnerability, either intentionally or unintentionally. If a cybersecurity incident occurs, there is a consequence to the WWU.

Cyber incidents can occur for many of the following reasons:

- Amateur hackers, with access to many online tools and resources, can find systems connected to the Internet and interfere with their operation.
- Professional hackers, with more skills and resources, can target organizations with ransomware and other disruptive techniques for profit.
- Activist hackers can disrupt the operations of organizations that they consider are not aligned with their political or social views.
- Disgruntled employees or contractors can use inside knowledge, or privileged access, to gain revenge.
- Nation-states or terrorists can target critical infrastructure organizations to create instability.

Cyber incidents need not be the result of deliberate malicious intent. Employees or contractors can also cause serious consequences, for instance, by simply failing to maintain adequate backups.

Vulnerabilities are flaws. They can be one of three types:

- Technical—a flaw in software or hardware, a configuration error

- Process—a failure in a process or procedure
- People—a failure to meet personal responsibilities

The next chapter will describe the process for WWUs to follow to understand and manage their cybersecurity risk, considering threats, vulnerabilities, and exploits.

Cybersecurity risk management is an essential part of WWU operations. This chapter focuses on a recommended approach that suits the unique constraints of WWUs, specifically limited budgets, and sparse in-house technical capability.

This approach is based on the methodology described in ISA/IEC62443-3-2, *Security Risk Assessment for System Design*. This document is part of the ISA/IEC62443 series of standards called *Security for Industrial Automation and Control Systems*.[43] The standards series is produced by the International Society of Automation (ISA), a standards development organization based in Raleigh, North Carolina. ISA/IEC62443 is the only internationally available consensus-based series of standards covering the systems and technology that WWUs depend on.

The approach described in this chapter has been distilled from the ISA/IEC62443-3-2 methodology so that it is focused on the needs and capabilities of WWUs. Further details of the series of standards are provided in Chapter 8, "Terminology, Resources, and Further Reading."

[43]International Society of Automation. (n.d.). *ISA/IEC-62443-3-2-2020, Security for industrial automation and control systems part 3-2: Security risk assessment for system design.*

Chapter 3

Assessing and Managing Cybersecurity Risk

1.0 INTRODUCTION

Cybersecurity risk management is an essential part of water and wastewater utility (WWU) operations. This chapter focuses on a recommended approach that can be used by WWUs of all sizes, including very small systems with their unique constraints of limited budgets and sparse in-house technical capability to help prioritize their mitigation plans.

This approach is based on the requirements described in the ISA/IEC 62443 series of standards called *Security for Industrial Automation and*

Control Systems.[1] The standards series is produced by the International Society of Automation (ISA), a standards development organization based in Raleigh, North Carolina. ISA/IEC 62443 is the only internationally available consensus-based series of standards covering the systems and technology that WWUs depend on.

Further details of the series of standards are provided in Chapter 8, "Terminology, Resources, and Further Reading."

2.0 RISK

2.1 Definition

Risk is "the effect of uncertainty on objectives" and risk management is the "coordinated set of activities and methods that is used to direct an organization and to control the many risks that can affect its ability to achieve objectives."[2]

Water and wastewater utilities are exposed to a wide variety of risks. At a very high level, they include

- contamination of water supply,
- loss of service,
- pollution of the environment,
- harm to employees and the public, and
- damage to equipment.

Water and wastewater utilities are required to identify their risks and manage them. America's Water Infrastructure Act identifies two key requirements: (1) completion of a risk assessment and (2) completion of an emergency response plan.

The Occupational Safety and Health Administration defines risk as "the product of hazard and exposure. Thus, risk can be reduced by controlling or eliminating the hazard or by reducing workers' exposure to hazards. An assessment of risk helps employers understand hazards in the context of their own workplace and prioritize hazards for permanent control."[3]

---

[1]International Society of Automation. (n.d.). *ISA/IEC-62443-3-2-2020, Security for industrial automation and control systems part 3-2: Security risk assessment for system design.*

[2]International Standards Organization. (n.d.). *ISO 31000:2018(E), Risk management—Guidelines* (page 1).

[3]Occupational Safety and Health Administration. (n.d.). *Recommended practices for safety and health programs.* Retrieved January 21, 2024, from https://www.osha.gov/safety-management/hazard-Identification

This definition can be repurposed to cover all risks for WWUs. Risk is the result of exposure to a hazard or threat. A hazard or threat could be environmental-based, such as flooding or an earthquake, or infrastructure-based, such as loss of power or communications, or others, such as cybersecurity-related. If left unmanaged, hazards or threats can result in incidents that have consequences. For example, flooding could render a facility inoperable (the incident), which, in turn, could mean a disruption of service for customers (the consequence).

Risk management focuses on reducing the risk by

- reducing the likelihood of the hazard causing an incident and
- reducing the consequences if the incident occurs.

2.2 Quantifying Risk

Because there are so many possible risks for a WWU, it is essential that they are quantified. Quantifying risks involves assessing the likelihood and consequence and allows WWUs to understand the risks they face and where they need to focus their resources to minimize these risks.

Figure 3.1 shows a sample risk matrix that can be used to quantify WWU risks.

Likelihood and consequence are categorized using four levels. For likelihood, the categorization ranges from improbable events (lowest likelihood) to frequent (highest likelihood). For consequence, the categories are "Low," "Moderate," "High," and "Severe." These are further broken down in this example into types of consequence:

- health and safety—harm to employees;
- environmental—damage to the environment;
- financial—additional costs for new equipment, employee time, and so forth;
- reputation and public disruption—the result of public or media response to an incident; and
- regulatory—fines or punishment from governmental bodies.

The specific details of each consequence may vary between different WWUs. For example, a larger WWU may be exposed to more financial risk.

Once risks are quantified, they will appear on the risk matrix in one of three areas:

- Green—tolerable risks that require limited active monitoring.
- Yellow—tolerable risks that require active monitoring.

		Consequence					Likelihood			
							Improbable (> 10 years)	Remote (Once in 10 years)	Occasional (Once a year)	Frequent (Once a month)
		Health & Safety	Environmental	Financial	Reputation & Public Disruption	Regulatory	1	2	3	4
Low	1	Injury or illness without a loss of a workday; Event requiring First Aid	Unfavorable impact but without any environmental harm or nuisance	Less than $5k	Public not aware	Limited impact with less than $10k in fines or penalties	1	2	3	4
Moderate	2	Injury or illness resulting in treatment by a doctor or hospitalization; Long term sickness over 3 days	Unreasonable impact resulting in environmental nuisance/cleanup	Between $5k and $50k	Local media coverage; Telephone complaints	Fines or penalties between $10k and $200k	2	4	6	8
High	3	Permanent injury or long-term sickness over 3 months; Multiple serious injuries	Significant impact with localized environmental harm but not in an environmentally sensitive area	Between $50k and $100k	Extended local adverse media campaign; State media coverage	Fines or penalties totaling between $200k and $500k	3	6	9	12
Severe	4	Single or multiple deaths; Multiple permanent injuries	Major environmental harm caused in environmentally sensitive areas; Environmental damage causing a violation of law	Over $100k	National adverse media coverage	Officer jailed Fine or penalty over $500k	4	8	12	16

FIGURE 3.1 Example Risk Matrix

- Red—intolerable risks that must be further managed to reduce the likelihood and/or consequence until it is tolerable.

2.3 As Low as Reasonably Practicable

Not all risks are equal. To make measurable progress, there needs to be a way to focus effort for maximum effect. Figure 3.2 visualizes this. The tolerable versus intolerable risk areas are shown, along with the relative cost for addressing these risks.

In general, the cost of addressing risk is inversely proportional. It is possible to address many intolerable risks for relatively low cost (e.g., ensuring that computer equipment is kept up to date with software fixes). But there is a point at which the cost of addressing risks becomes prohibitive or any cost spent to address improbable risks is not justifiable. This point is called *as low as reasonably practicable* or ALARP for short.

ALARP should not be used as a justification for not addressing risks. It should, however, be used to help WWUs focus on addressing the intolerable risks they have that require the least investment. This prioritization will help overcome one of the challenges with cybersecurity risk management: where to focus efforts for best effect. There are many examples of the use

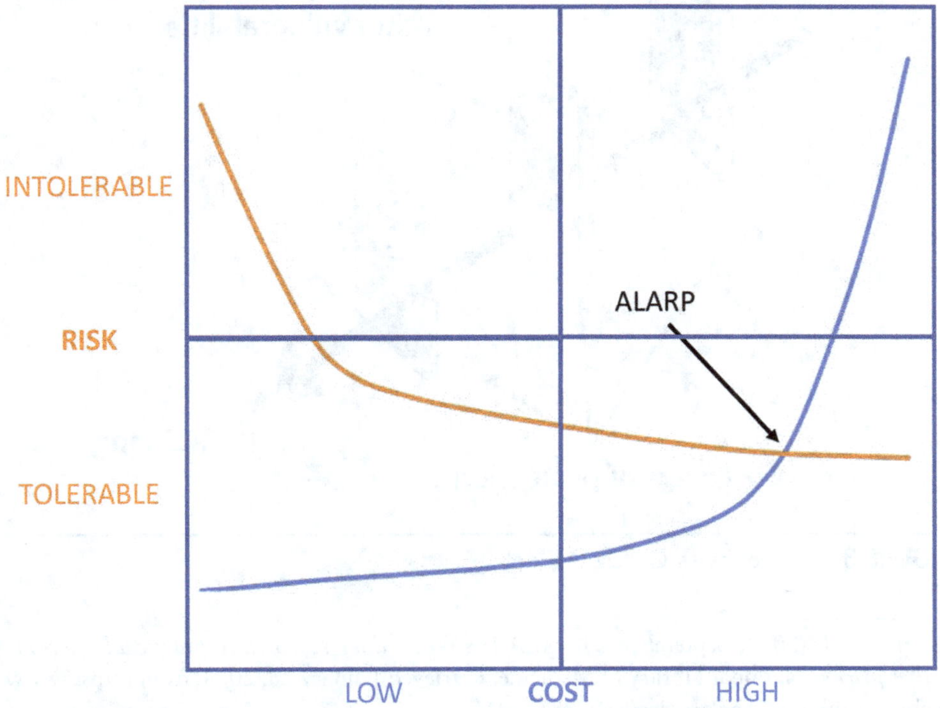

FIGURE 3.2 As Low as Reasonably Practicable

of ALARP across sectors. In one example from the Healthcare and Public Health Sector, researchers concluded that "a cost–benefit calculation based on the legal definition of 'as low as reasonably practicable' may have some use in the support of radiation protection decision-making in the hospital environment, but only within the context of existing legal, practical and ethical considerations."[4]

2.4 Addressing Cybersecurity Risk

In risk management, reducing likelihood and consequence is achieved by introducing *barriers*. This barrier concept is used globally across sectors as a means of understanding how risks are being managed. Burlingame and Bartrand use the approach to describe how water safety plans reduce public health risk.[5] In cybersecurity, this barrier concept is known as *defense in depth*.

The barrier concept was created by Professor James Reason of the University of Manchester Institute of Technology in 2000. He explained that systems that include methods to reduce risk also have inherent flaws, and when these flaws are aligned, accidents can occur. Figure 3.3 shows a version

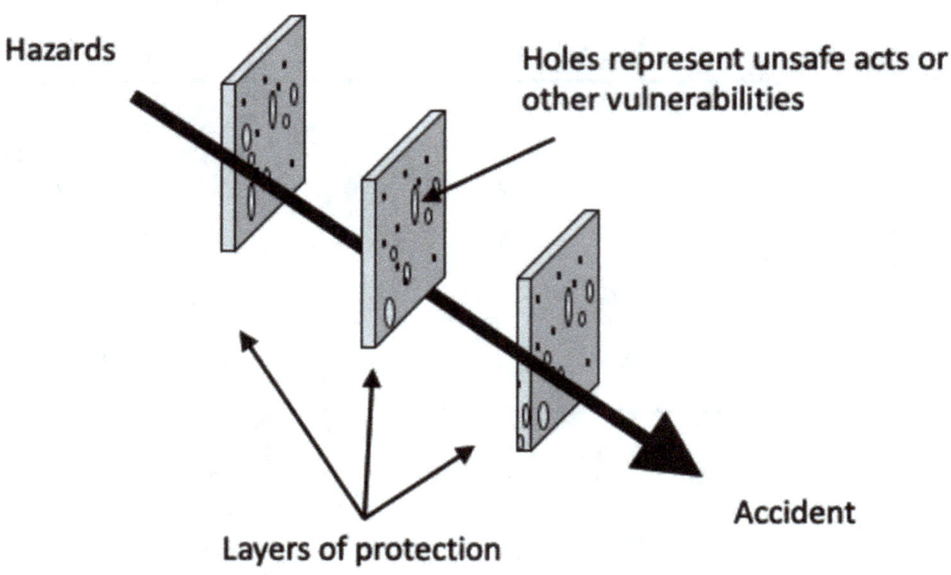

FIGURE 3.3 The Swiss Cheese Model

[4]Kotre, C. J. (2022, September 20). *ALARP: When does reasonably practicable become rather pricey?* National Library of Medicine. Retrieved January 28, 2024, from https://www.ncbi.nlm.nih.gov/pmc/articles/PMC9815731/

[5]Burlingame, G. A., & Bartrand, T. (2023, June). How water safety plans complement sanitary surveys. *AWWA Journal, 115*(4), 44–56.

of his 1990 drawing[6] representing the barriers as slices of Swiss cheese. The slices of cheese represent the barriers, and the holes in the slices of cheese represent the flaws. This concept has also become known as the *Swiss cheese model.* There are other risk models that can be used. This book will focus on the Swiss cheese model as it helps WWUs visualize the controls that must be in place to manage cybersecurity.

A cybersecurity barrier model for WWUs is shown in Figure 3.4. Each barrier identifies a specific technical or procedural control that WWUs should have in place to manage their cybersecurity risk.

The barrier model in Figure 3.4 has two distinct types of barriers:

- prevention—these barriers are required to reduce the likelihood of a cybersecurity incident occurring and
- mitigation—these barriers are required to reduce the consequences of a cybersecurity incident when it does occur.

Water and wastewater utilities can use this model as part of a cybersecurity risk management process that will help identify, quantify, and manage their risks.

3.0 CYBERSECURITY RISK MANAGEMENT PROCESS

3.1 Overview

As noted in the introduction the approach described in this book, including the risk management process, is based on the requirements described in the ISA/IEC 62443 series of standards. Although this series of standards is aimed at industrial automation and control systems (or IACS, e.g., WWU supervisory control and data acquisition [SCADA]), the approach can also cover the information technology requirements in a typical WWU.

The ISA/IEC 62443 series of standards defines requirements for

- an overall security program for an organization,
- a risk assessment process,
- system security,
- product development life cycle processes, and
- system component security.[7]

[6]Reason, J. (1990). *Human error* (p. 208). Cambridge University Press.

[7] International Society of Automation. (n.d.). *ISA/IEC 62443*. Retrieved November 11, 2023, from https://isagca.org/isa-iec-62443-standards

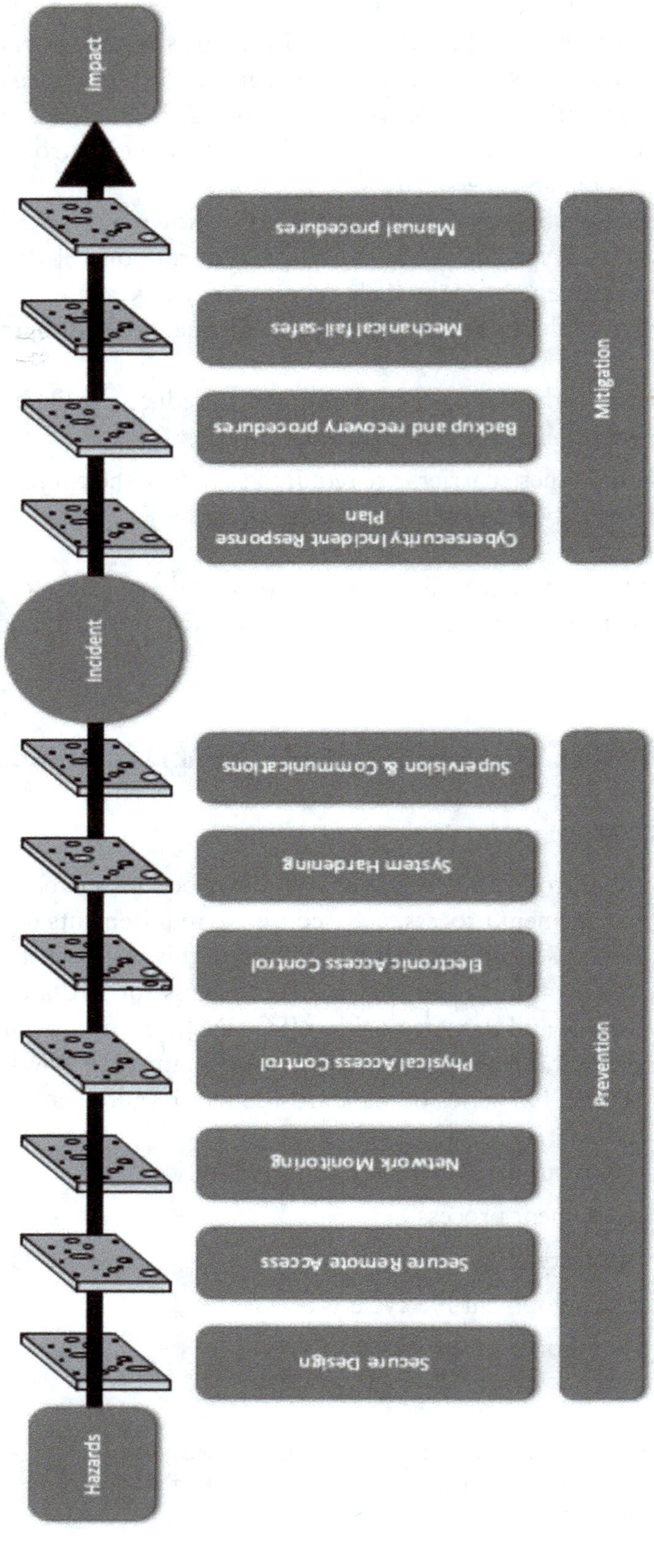

FIGURE 3.4 Cybersecurity Barrier Model for WWUs

The requirements for the security program and for system security have been incorporated to create the cybersecurity barrier model described earlier in this chapter. For systems security, the series of standards defines four distinct security levels (SL), as shown in Table 3.1, each dealing with protection against an increasing level of risk.

This methodology focuses on helping WWUs achieve SL 1 throughout their operational environment, with some SL 2, SL 3, and SL 4 capabilities where possible (e.g., use of multifactor authentication in user authentication and authorization). The methodology can be used in conjunction with the series of standards to identify the specific system requirements needed to achieve a particular security level.

Water and wastewater utilities should apply the product development life cycle process and system component security requirements to product vendors. This is described in Chapter 6, "Dealing With Third Parties."

The requirements for risk assessment have been incorporated into the process shown in Figure 3.5. There are four stages:

1. High-level risk assessment—to identify and quantify risk.
2. Gap assessment—to identify the vulnerabilities in the cybersecurity barrier model.
3. Remediation—to correct the vulnerabilities identified in the gap assessment.
4. Maintenance—to ensure the cybersecurity barriers are maintained, and the WWU is prepared to manage a potential cybersecurity incident.

3.2 High-Level Risk Assessment

This stage is conducted first, to identify and quantify the risks to the WWU.

TABLE 3.1 ISA/IEC62443 Security Level Definitions

SL 1	Protection against casual or coincidental violation
SL 2	Protection against intentional violation using simple means with low resources, generic skills, and low motivation
SL 4	Protection against intentional violation using sophisticated means with moderate resources, IACS specific skills, and moderate motivation
SL 4	Protection against intentional violation using sophisticated means with extended resources, IACS specific skills, and high motivation

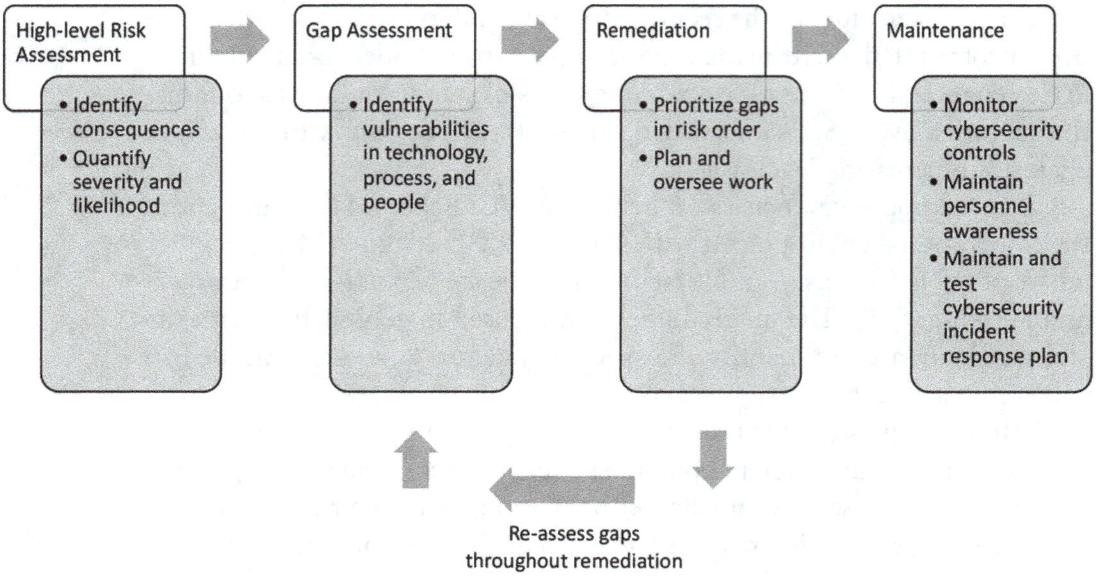

FIGURE 3.5 Cybersecurity Risk Management Process

A workshop should be conducted with all key individuals and experts in the WWU. The exact list of attendees will depend on the specific organization, but should include representatives who understand

- operational issues,
- technical issues, and
- business or financial issues.

The workshop should be facilitated by someone from outside the WWU. This helps avoid unintentional bias to occur and will help ensure all viewpoints are heard equally.

The objective of the workshop is to agree on all operational activities of a WWU and how they could be affected by a cybersecurity incident. No incident should be dismissed regardless of how unlikely. This risk assessment helps the WWU team understand where the biggest risks exist and, therefore, what to focus efforts on. This risk assessment will also be used to develop the cybersecurity incident response plan. The result of this stage should be a completed risk matrix with notes explaining the details of how the quantification was determined.

3.3 Gap Assessment

3.3.1 Questionnaires

This stage determines whether cybersecurity barriers are in place and helps identify a list of gaps that need to be addressed.

The stage uses a questionnaire to guide the assessment. There are multiple forms of questionnaires that exist for WWUs to use. These are described in more detail in Chapter 8, "Terminology, Resources, and Further Reading."

Care must be taken when using assessment questionnaires. The assessment needs to focus on the issues that are appropriate to WWUs, but also cover all issues relating to cybersecurity.

Some questionnaires are so long and complex that it is difficult to identify the critical issues from the results. Some questionnaires miss essential barriers such as physical access or removable media control, giving a false impression that these are not required.

For WWUs, any questionnaire should address the barriers identified in Figure 3.4 and detailed in Table 3.2.

The gap assessment should also involve a workshop with all key individuals and experts in the WWU. The list of attendees is likely to be the same as that for the high-level risk assessment workshop.

As before, the workshop should be facilitated by someone from outside the WWU to avoid demonstrating unintentional biasness; thus, ensuring that all viewpoints are heard equally. It is also important to have someone independently facilitate the effort to ensure that all gaps are captured. Self-assessment carries the risk that some gaps may be overlooked to avoid creating additional work on already overloaded employees.

3.3.2 Vulnerability Review

Product vulnerabilities are now tracked in a variety of publicly available databases. Each vulnerability is assigned a unique identifier that can be used to search directly for information related to that vulnerability (for instance, software patches that remedy the vulnerability).

The National Institute of Standards and Technology maintains the National Vulnerability Database, which is a valuable resource for finding vulnerabilities in information technology (IT) and operational technology (OT) equipment.[8] This database is synchronized from CVE, the Common

[8]National Institute of Standards and Technology. (n.d.). *National vulnerability database.* Retrieved February 11, 2024, from https://nvd.nist.gov/

TABLE 3.2 WWU Cybersecurity Barrier Definitions

Prevention Barriers	Secure design	The technical controls to separate key elements of the WWU infrastructure and protect against unauthorized access from outside the organization.
	Secure remote access	The technical and procedural controls that are involved with authorized users accessing WWU resources from outside of the organization.
	Network monitoring	The systems and procedures in place to detect potential cybersecurity incidents.
	Physical access control	The systems and procedures in place to limit physical access to WWU facilities.
	Electronic access control	The systems and procedures in place to limit electronic access to WWU facilities.
	System hardening	The systems and procedures in place to manage the security of programmable electronic equipment such as personal computers and PLCs. This includes anti-malware; software updates; and management of universal serial bus (USB) drives, laptops, and other devices.
	Supervision and communications	The various processes and procedures that are in place to maintain secure and safe operations. This includes training and personnel processes and covers employees as well as third parties such as vendors and contractors.
Mitigation Barriers	Cybersecurity incident response plan	The level of preparedness that the WWU has for dealing with cybersecurity incidents, including quality of plan and frequency of testing.
	Backup and recovery procedures	The processes in place for backing up electronic programmable equipment and testing recovery periodically.
	Mechanical fail-safes	Anything that is not electronic or programmable that is used to mitigate an issue. Examples include pressure relief systems on filter equipment and chemical containment devices. Electronic or programmable equipment is excluded as it can be subject to a cybersecurity incident.
	Manual procedures	Documented procedures that may be required to be executed in the event of a cybersecurity incident. For example, operating a water treatment facility without automatic control equipment.

Vulnerabilities and Exposures database, maintained by the MITRE Corporation. The MITRE Corporation was founded in 1958 as a not-for-profit company to serve as advisers in systems engineering to government agencies, both military and civilian.[9]

The ICS Advisory Project provides smaller OT asset owners, chief information security officers (CISOs), cybersecurity analysts, and researchers with an analysis tool that allows them to quickly identify threats and vulnerabilities by product, vendor, and by Critical Infrastructure sector. The data for the ICS Advisory Project are sourced from publicly available threat and vulnerability data such as CVE, but are focused on OT equipment, such as that used in the SCADA systems used by WWUs.[10]

These databases allow searches for product vulnerabilities. Searches can be conducted by product name, vendor, vulnerability identifier, or other relevant criteria. An example showing a vulnerability in Microsoft Outlook, taken from the MITRE CVE website, is shown in Figure 3.6.

The purpose of this exercise is to determine what vulnerabilities exist so that action can be taken, as follows:

- Search for product updates that remediate the vulnerabilities. The vulnerability databases will usually link to an update from the vendor, if available.
- Take alternative actions to mitigate a vulnerability for which there is no product update available. The vulnerability databases will usually provide temporary remediation that can be used until the vulnerability has been resolved by a vendor patch.
- Apply approved vendor security patches as soon as is practical. Note that care must be taken to verify that one vendor's update does not affect another's. For example, a Windows update may be incompatible with SCADA software running on the machine.

It is recommended that WWUs stay informed about newly discovered vulnerabilities by subscribing to email alerts from CISA, MITRE, or other relevant sources.

3.3.3 Other Gap Assessment Methods

Other methods can be used in the gap assessment process, in addition to questionnaires and vulnerability reviews. Table 3.3 provides the most common methods.

---

[9]Mitre. (n.d.). *CVE program mission*. Retrieved February 11, 2024, from https://www.cve.org/

[10]ICS Advisory Project. (n.d.). *ICS advisory project*. Retrieved February 11, 2024, from https://www.icsadvisoryproject.com/

CVE-ID

CVE-2023-36895 Learn more at National Vulnerability Database (NVD)
• CVSS Severity Rating • Fix Information • Vulnerable Software Versions • SCAP Mappings • CPE Information

Description

Microsoft Outlook Remote Code Execution Vulnerability

References

Note: References are provided for the convenience of the reader to help distinguish between vulnerabilities. The list is not intended to be complete.

- MISC:Microsoft Outlook Remote Code Execution Vulnerability
- URL:https://msrc.microsoft.com/update-guide/vulnerability/CVE-2023-36895

Assigning CNA

Microsoft Corporation

Date Record Created

20230627 Disclaimer: The record creation date may reflect when the CVE ID was allocated or reserved, and does not necessarily indicate when this vulnerability was discovered, shared with the affected vendor, publicly disclosed, or updated in CVE.

Phase (Legacy)

Assigned (20230627)

Votes (Legacy)

Comments (Legacy)

Proposed (Legacy)

N/A

This is an record on the CVE List, which provides common identifiers for publicly known cybersecurity vulnerabilities.

SEARCH CVE USING KEYWORDS: Submit

You can also search by reference using the CVE Reference Maps.

For More Information: CVE Request Web Form (select "Other" from dropdown)

FIGURE 3.6 Vulnerability Example From CVE

TABLE 3.3 Common Cybersecurity Gap Assessment Methods

Walkthrough	Can yield additional information on top of workshop through: • Visual inspection of the system • Comparison of drawings with actual installation • Validation of physical security (doors, locks, etc.) • Observation of behavior in operating environment • Interviews with operational personnel
Passive	This may result in discovery of technical vulnerabilities through: • Capture and study actual network traffic • Collection of log data from devices • Review of device configurations
Active	Sometimes called a *penetration*, or *pen test*, this may result in discovery of technical vulnerabilities using active network scanning tools and techniques. This presents a greater risk than the walkthrough and passive methods and so should be used with great care.
Red Team Assessment	This method gets its name from military wargaming, in which conflicts are simulated between an aggressor, the red team, and a defending force, the blue team. Red-team assessments in cybersecurity involve experts attempting to achieve a target, such as access to a certain machine or other resource. The exercise uses realistic scenarios that can identify vulnerabilities in technology, people, or processes.

3.3.4 Results of Gap Assessment

These results will form the basis of a remediation plan, and so it is essential that they are clearly documented. The result of the stage should be a report listing all the gaps. This could be in the form of a spreadsheet or document.

3.4 Remediation

With the results of the gap assessment, the WWU should prioritize the remediation of the gaps based on a combination of risk exposure and cost. High-risk/low-cost remediation should be the priority, followed by moderate risk/moderate cost.

Some gaps may take longer to remediate than others, in some cases weeks or months. For this reason, it is important to have an overall plan showing the remediation stages. The gap assessment should be used to track progress and risk reduction during this remediation phase.

3.5 Maintenance

The maintenance phase should start as soon as the gap assessment is complete. The purpose of the maintenance phase is to ensure the cybersecurity barriers are in place and effective. Even though some barriers may not be in place until remediation is complete, others will be, and it is essential that all barriers are always maintained.

The other key element of the maintenance phase is to ensure the cybersecurity incident response plan is in place and regularly tested. The WWU should choose various scenarios and perform desk-based and realistic on-site exercises to ensure the response plan will work if needed.

4.0 EXAMPLE

4.1 System Description

Consider this as an example *reference WWU* that incorporates typical features for most WWUs. Figure 3.7 shows an overview of the reference system.

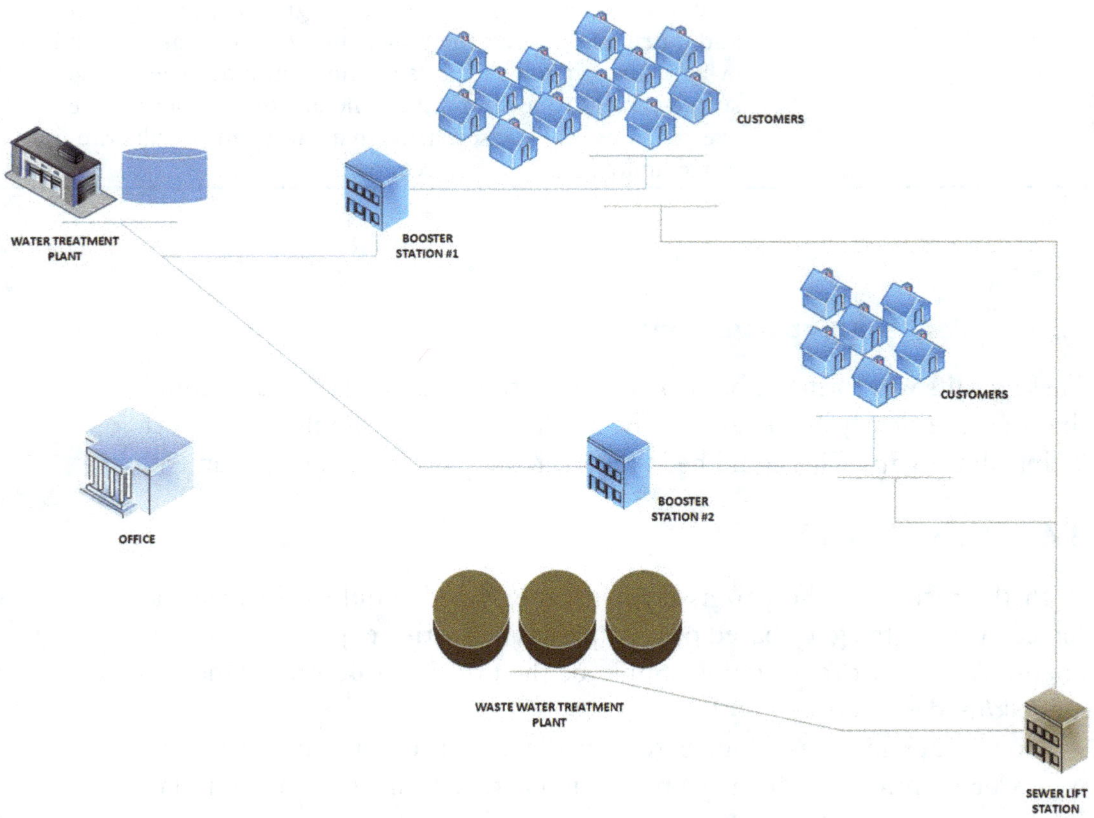

FIGURE 3.7 Reference WWU

The reference system incorporates

- a water treatment facility,
- two remote booster stations,
- a water resource recovery facility (WRRF),
- a remote sewer lift station, and
- an office.

This system contains a variety of technologies used to support operational activities. Figure 3.8 shows an overview of the key assets and their connectivity.

The main technology consists of four main systems:

- A SCADA system, used to monitor and control the physical processes at all the sites;

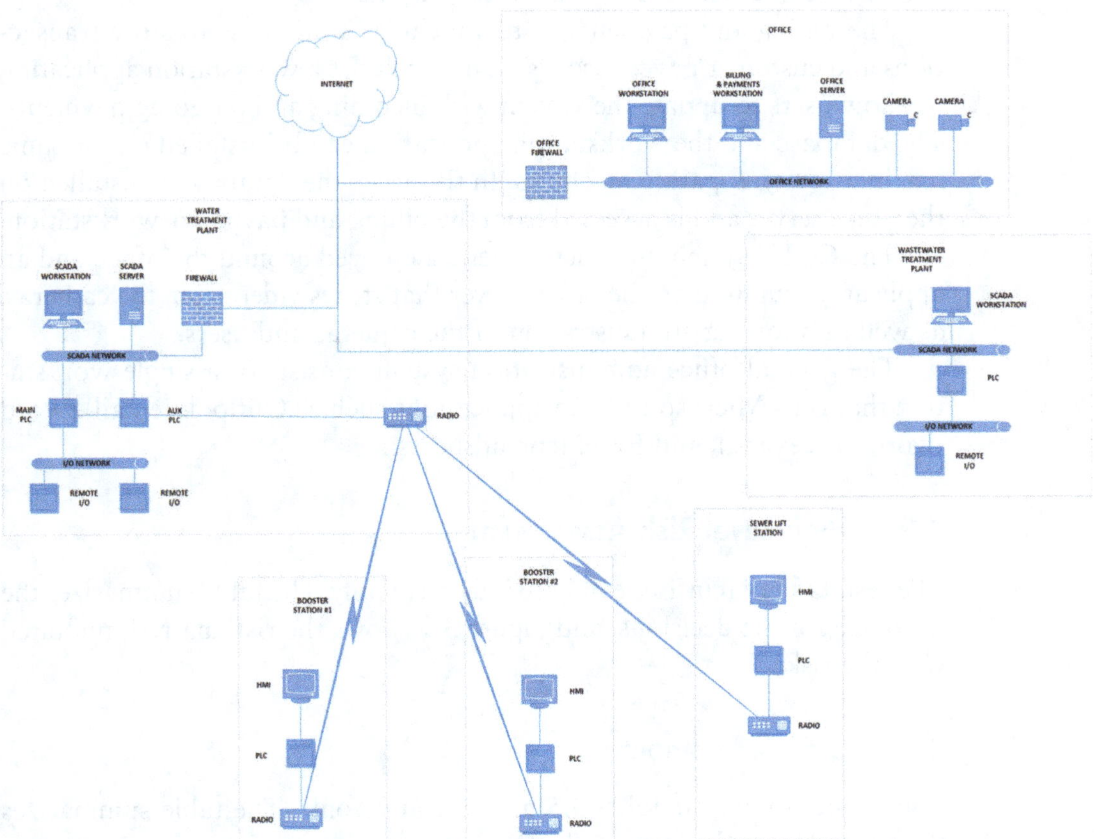

FIGURE 3.8 Technology Overview of Reference WWU

- A billing and payments system, used to bill customers and pay employees and suppliers. This system may include smart meter technology deployed at customer locations;
- A closed-circuit television (CCTV) system; and
- A general office administration system.

Other systems may be present in WWUs. Common examples are a Geospatial Information System and a Computerized Management System. The assessment process applied to the example WWU would likely identify similar issues in these systems.

The SCADA system includes devices at the two remote booster stations and the sewer lift station called *programmable logic controllers* (PLCs) that read process data from sensors and control devices such as valves and motors.

Information is displayed on a local panel called a *human–machine interface*, which may communicate back to the main SCADA site via a direct radio link. Users at the main SCADA site can view data from all remote sites, as well as send commands to remote sites.

The billing and payments system includes a database to store transactions and customer/payee records. It also includes a workstation application to allow users to update the system with incoming and outgoing payments. The database and the workstation application can be installed on the same machine, or on separate machines. In this case, the database is installed on the office server and is accessed from the billing and payments workstation.

The CCTV system includes cameras deployed around the office and an application running on the office server that stores video from the cameras, as well as allowing administration of the cameras and users.

The general office administration system consists of a single workstation that runs Microsoft Office applications such as Outlook (email), Word (word processing), and Excel (spreadsheets).

4.2 High-Level Risk Assessment

The results shown in Table 3.4 are indicative only. The table summarizes the rationale for the decisions, and Figure 3.9 shows the risk matrix annotated with the risks.

4.3 Gap Assessment

The results shown in Table 3.5 are indicative only. The table summarizes the gaps identified in the workshop.

TABLE 3.4 Indicative Results from the High-Level Risk Assessment

Risk ID	Consequence	Rationale
1	Loss of Water Treatment Facility	It will be necessary to provide other sources of potable water for customers while the water treatment facility is out of service. This may require invoking emergency measures to obtain water from a neighboring system or purchasing bottled water. This will incur significant costs and is not sustainable for long periods (more than 1 or 2 days). Loss of water treatment facility is a remote, but not improbable scenario.
2	Loss of Booster Station #1	Booster station #1 covers a larger population than station #2 so the effect is greater. Lack of booster station creates similar results to loss of the water treatment facility, but as its scope is smaller, it can be tolerated for slightly longer periods. Water from another part of the network can be used in the short term, for example. Loss of the complete booster station is improbable because it has a dual pump system so both pumps would need to fail simultaneously.
3	Loss of Booster Station #2	As noted for booster station #1, the effect for the loss of booster station #2 is smaller. Water from another part of the network can be used in the short term. Loss of the complete booster station is improbable because it has a dual pump system so both pumps would need to fail simultaneously.
4	Loss of Sewer Lift Station	Loss of sewer lift station will cause wastewater to back up in the network. This will need to be resolved using temporary pump equipment, involving significant cost and employee time. Loss of the complete lift station is remote because it has a dual pump system so both pumps would need to fail simultaneously, but it is more likely than the booster station to fail due to the nature of the medium being pumped.
5	Loss of Office	Loss of the office could have various negative effects depending on level of preparedness. If the billing and payment system can be restored to another location, then most activities can be maintained. Without the billing and payment system, there would need to be manual processes for collections and payments. In this assessment, the worst case is assumed. Manual procedures can run for up to 30 days, after which it becomes more difficult to keep records. It is assumed that loss of office function could occur occasionally due to a variety of reasons (e.g., fire, theft, etc.).

							Likelihood			
Consequence							Improbable (> 10 years)	Remote (Once in 10 years)	Occasional (Once a year)	Frequent (Once a month)
		Health & Safety	Environmental	Financial	Reputation & Public Disruption	Regulatory	1	2	3	4
Low	1	Injury or illness without a loss of a workday; Event requiring First Aid	Unfavorable impact but without any environmental harm or nuisance	Less than $5k	Public not aware	Limited impact with less than $10k in fines or penalties	1	2	3	4
Moderate	2	Injury or illness resulting in treatment by a doctor or hospitalization; Long term sickness over 3 days	Unreasonable impact resulting in environmental nuisance/cleanup	Between $5k and $50k	Local media coverage; Telephone complaints	Fines or penalties between $10k and $200k	2	4	❺ 6	8
High	3	Permanent injury or long-term sickness over 3 months; Multiple serious injuries	Significant impact with localized environmental harm but not in an environmentally sensitive area	Between $50k and $100k	Extended local adverse media campaign; State media coverage	Fines or penalties totaling between $200k and $500k	❸ 3	❹ 6	9	12
Severe	4	Single or multiple deaths; Multiple permanent injuries	Major environmental harm caused in environmentally sensitive areas; Environmental damage causing a violation of law	Over $100k	National adverse media coverage	Officer jailed Fine or penalty over $500k	❷ 4	❶ 8	12	16

FIGURE 3.9 High-level Risk Assessment Matrix

TABLE 3.5 Indicative Results of Gap Assessment

Prevention Barriers	Secure design	1. There is no firewall at the WRRF.
		2. The Office firewall does not limit access to the network from the Internet.
		3. The Office server has the billing and payments database and CCTV camera system deployed on it.
		4. The SCADA system does not limit setpoints to safe levels.
	Secure remote access	5. The Office workstation allows remote access to the SCADA system server at the water treatment facility.
		6. The SCADA vendor has uncontrolled remote access to the SCADA system server at the water treatment facility (e.g., dial-up modem).
	Network monitoring	7. There is no monitoring of incoming traffic from the Internet on the Office firewall.
		8. There is no procedure to allow employees to report cybersecurity incidents.
	Physical access control	9. Cellular operator has access to booster station #1 building as cell tower is on the same site.
	Electronic access control	10. The SCADA system has shared user accounts used by employees and vendors.
		11. Office workstation has no electronic access control.
		12. There is no multifactor authentication required to access any system.
		13. The PLCs and SCADA systems have default accounts or passwords.
	System hardening	14. The USB ports on all computers are open for anyone to access.
		15. The PLCs have a webserver enabled.
		16. Network ports are physically unsecured.
	Supervision and communications	17. There is no joiners/movers/leavers process in place and, as a result, leavers may still have physical and electronic access to equipment.
		18. There is no cybersecurity training for employees or third parties.

(continued)

TABLE 3.5 Indicative Results of Gap Assessment

Prevention Barriers (*continued*)	Cybersecurity incident response plan	19. The current plan is a generic plan that does not cover the scenarios identified in the high-level risk assessment.
		20. The current plan has out-of-date contact information.
		21. The plan is not held and maintained at each physical location.
		22. There is no process for periodically testing the plan or training employees in that plan.
Mitigation Barriers	Backup and recovery procedures	23. There are no backups of the PLC programs other than with the System Integrator.
		24. There are no backups of the PLC's configuration files other than with the System Integrator.
		25. Backups of billing and payment system are held on storage device in Office.
	Mechanical fail-safes	26. No gaps identified
	Manual procedures	27. There is no written procedure for manually operating the water treatment facility and no plan for training employees in that procedure.

4.4 Prioritization for Remediation

For the purposes of this example, the gaps have been prioritized in Table 3.6. The following three priorities have been defined:

- Priority 1—These vulnerabilities expose the WWU to the highest levels of risk and need to be addressed within a month.
- Priority 2—These vulnerabilities are severe, but resources need to address priority 1 gaps first. These need to be addressed within 3 months.
- Priority 3—These vulnerabilities are to be addressed when priority 1 and 2 gaps are closed. These need to be addressed within 6 months.

The highest risk arises from the following items:

- Lack of firewall between office and WRRF, and uncontrolled access to the office through the office firewall. Correcting the office firewall and adding a firewall with appropriate rules between the office and WRRF are relatively simple tasks.

TABLE 3.6 Prioritized List of Gaps

Gap	Priority
1. There is no firewall at the WRRF.	1
2. The Office firewall does not limit access to the network from the Internet.	1
3. The Office server has the billing and payments database and CCTV camera system deployed on it.	3
4. The SCADA system does not limit setpoints to safe levels.	2
5. The Office workstation allows remote access to the SCADA system server at the water treatment facility.	1
6. The SCADA vendor has uncontrolled remote access to the SCADA system server at the water treatment facility.	1
7. There is no monitoring of incoming traffic from the Internet on the Office firewall.	3
8. There is no procedure to allow employees to report cybersecurity incidents.	3
9. Cellular operator has access to booster station #1 building as cell tower is on the same site.	3
10. The SCADA system has shared user accounts used by employees and vendors.	3
11. Office workstation has no electronic access control.	2
12. There is no multifactor authentication required to access any system.	2
13. The PLCs and SCADA systems have default accounts or passwords.	2
14. The USB ports on all computers are open for anyone to access.	2
15. The PLCs have webserver enabled.	3
16. Network ports are physically unsecured.	3
17. There is no joiners/movers/leavers process in place and, as a result, leavers may still have physical and electronic access to equipment.	1
18. There is no cybersecurity training for employees or third parties.	2
19. The current cybersecurity incident response plan is a generic plan that does not cover the scenarios identified in the high-level risk assessment.	1
20. The current cybersecurity incident response plan has out-of-date contact information.	1
21. The plan is not held and maintained at each physical location.	1
22. There is no process for periodically testing the cybersecurity incident response plan or training employees in that plan.	1
23. There are no backups of the PLC programs other than with the System Integrator.	2
24. There are no backups of the PLC's configuration files other than with the System Integrator.	2
25. Backups of billing and payment system are held on storage device in the office.	2
26. There is no written procedure for manually operating the water treatment facility and no plan for training employees in that procedure.	3

- Lack of reliable and tested cybersecurity incident response plan. This alone will make the difference between a service outage of a few hours and a service outage of several days or weeks, plus significant cost.
- Uncontrolled remote access to the SCADA system. If necessary, this should be removed until a secure solution can be identified. That includes access for the vendor.
- The lack of joiners/movers/leavers process. The issue is that when employees leave, they need to have their access (physical and electronic) removed immediately. This means recovering any keys, changing locks if necessary, and removing and changing usernames and passwords.

It is essential that these are identified as high-priority and addressed promptly. Diluting the list of tasks with other lower-priority issues could mean these are not addressed quickly enough.

5.0 SUMMARY

This chapter has provided an overview of cybersecurity risk management. The approach is based on the requirements described in the ISA/IEC62443 series of standards produced by the ISA. It is the only internationally available consensus-based series of standards covering the systems and technology on which WWUs rely.

A process based on this series of standards, suitable for WWUs to assess and manage their cybersecurity risks, has been provided. This process can be used by WWUs of all sizes, including very small systems with their unique constraints of limited budgets and sparse in-house technical capability.

An example of a reference WWU has been used to work through the assessment phase of the process, considering a high-level risk assessment, a gap assessment, and a prioritization of the gaps.

The next chapter will describe the most common cybersecurity controls in more detail and highlight some of the issues for WWUs to be aware of when implementing these controls.

Chapter 4

Cybersecurity Barriers

1.0 INTRODUCTION

Chapter 3, "Assessing and Managing Cybersecurity Risk," introduced the cybersecurity barrier model for water and wastewater utilities (WWUs). This is way to highlight all the barriers, or controls, that must be in place to provide good management of cybersecurity risks.

This chapter will consider each of the barriers in more detail, highlighting best practices and common issues.

2.0 THE CYBERSECURITY BARRIER MODEL

2.1 Overview

Figure 3.4 in Chapter 3, "Assessing and Managing Cybersecurity Risk," provides a visual representation of the cybersecurity barrier model for WWUs. This visual representation is very important as it can be used to support awareness training for employees, vendors, and contractors, highlighting all the controls that must be in place for secure and safe operations.

To better understand the concept of barriers, consider the example from Burlingame and Bartrand, describing how water safety plans reduce public health risk. They explain that "in order to deliver safe drinking water to customers, utilities must install, maintain, and operate multiple barriers to protect against chemical, radiological, and microbiological hazards that could threaten public health." They then describe the following barriers:

- protecting the source waters against contamination,
- water treatment such as filtration and disinfection,
- the use of disinfectant residual in distribution systems,
- maintenance of water mains,
- inspection of water storage facilities,
- water sampling and testing, and
- regulatory oversight.

"All of these are barriers or controls that work in concert to protect against water contamination or to mitigate harm should contamination occur."[1]

Although Burlingame and Bartrand did not separate the above barriers into prevention and mitigation categories, this can be done to further understand the purpose of each barrier:

- prevention—these barriers are required to reduce the likelihood of a cybersecurity incident occurring and
- mitigation—these barriers are required to reduce the consequences of a cybersecurity incident when it does occur.

One reason this distinction is important in cybersecurity risk management is to ensure that a WWU has both prevention *and* mitigation barriers in place, not just one or the other type. For example, having good electronic access control (prevention) is important, but if there is no cybersecurity incident response plan (mitigation) in place, then if the electronic access control barrier failed, the WWU may not be able to recover in a timely manner. Likewise, having a cybersecurity incident response plan in place is essential but it should be possible to prevent most cybersecurity incidents *before* the response plan is needed by having prevention barriers in place.

3.0 CYBERSECURITY PREVENTION BARRIERS

The following cybersecurity prevention barriers must be installed, maintained, and operated to reduce the likelihood of a cybersecurity incident, whether it be malicious or unintentional, from outside the WWU or from within:

- secure design,
- secure remote access,
- network monitoring,
- physical access control,
- electronic access control,
- system hardening, and
- supervision and communications.

[1]Burlingame, G. A., & Bartrand, T. (2023, June). How water safety plans complement sanitary surveys. *AWWA Journal, 115*(5), 44–56.

3.1 Secure Design

3.1.1 Network Security

Every WWU has some sort of network to manage. A small office with one or two workstations for billing and payments will have a connection to the Internet. Employees will use email to communicate, and this will involve the use of the Internet.

Water and wastewater utilities of all sizes will have some sort of control and monitoring equipment at remote sites, such as booster stations or lift stations. This equipment may be connected to the office to allow for remote monitoring of these sites.

One of the biggest exposures a WWU has is through these external connections. If equipment is connected to the Internet for valid reasons, then it is possible for malicious attackers to gain access through this route. *Firewalls*, if properly configured, can protect against this risk.

A common issue is to connect all equipment in a WWU together on the same network. This might include user workstations, billing and payment systems, supervisory control and data acquisition (SCADA) systems, meter reading systems, and so on. Although this can be convenient, it means that if one machine on the network is compromised, it is much easier to gain access to other systems. Physically separate networks, partitioned by firewalls or data diodes, can protect against this risk. For smaller organizations such as most WWUs, it may be more appropriate to use *virtual local area networks*, or *VLANs*.

3.1.1.1 Firewalls

Firewalls restrict network *traffic* (data) to or from a network or a part of a network. Any organization that has a connection to the Internet has a firewall, even if it is the one that is built into their Internet gateway device provided by their Internet service provider. Microsoft Windows includes a firewall application that can limit network traffic to and from the device it is running on.

Consider the example system described in Chapter 3, "Assessing and Managing Cybersecurity Risk." The office network is connected to the Internet via a firewall, as shown in Figure 4.1.

The purpose of this firewall is to limit what communications can take place between the office network and the Internet. By default, firewalls are configured to allow all traffic to and from any device. This is the situation in the reference WWU example. As noted in the vulnerability assessment in Chapter 3, the office firewall does not limit access to the network from the Internet.

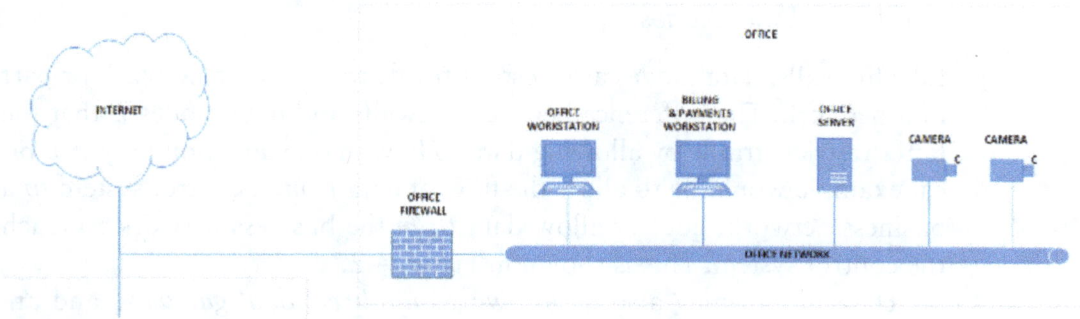

FIGURE 4.1 Firewall Protecting a WWU's Office's Network

To properly secure the office network, it will be necessary to

- identify what applications are running on the devices on the office network;
- create and maintain an inventory of all equipment on both office and plant networks;
- understand whether each application requires access to the Internet, and why;
- add rules to the firewall if the application needs access to the Internet; and
- configure the firewall to deny all other communications after all the rules are defined.

Firewall rule format will vary depending on the firewall, but at a minimum, each rule will define

- the source (which machine can send data),
- the destination (which machine can receive data), and
- what type of data can be sent between source and destination (the protocol).

It may be necessary to obtain help from product vendors to define these rules.

Firewalls can also log attempts to communicate with devices or use disallowed protocols. This logging is important and is discussed later in this chapter.

3.1.1.2 Data Diodes

Like firewalls, data diodes also restrict traffic to or from a network or part of a network. The difference between firewalls and data diodes is that the latter restricts traffic by allowing data to flow in one direction only. A typical example would be to allow the flow of data *from* a control system *to* a business network, but not allow data *from* the business network *to* reach the control system. This is shown in Figure 4.2.

Data diodes may also be known as *unidirectional gateways* and are most often used in sectors such as defense, government, finance, and energy.

Unlike firewalls, which use software and rules to define the flow of traffic, the unidirectional communications rules in a data diode are implemented in hardware making it inherently difficult to reverse.

3.1.1.3 Virtual Local Area Network

Considering the office network shown in Figure 4.1, the office workstation, billing and payments workstation, office server, and closed-circuit television (CCTV) cameras are on the same local area network (LAN).

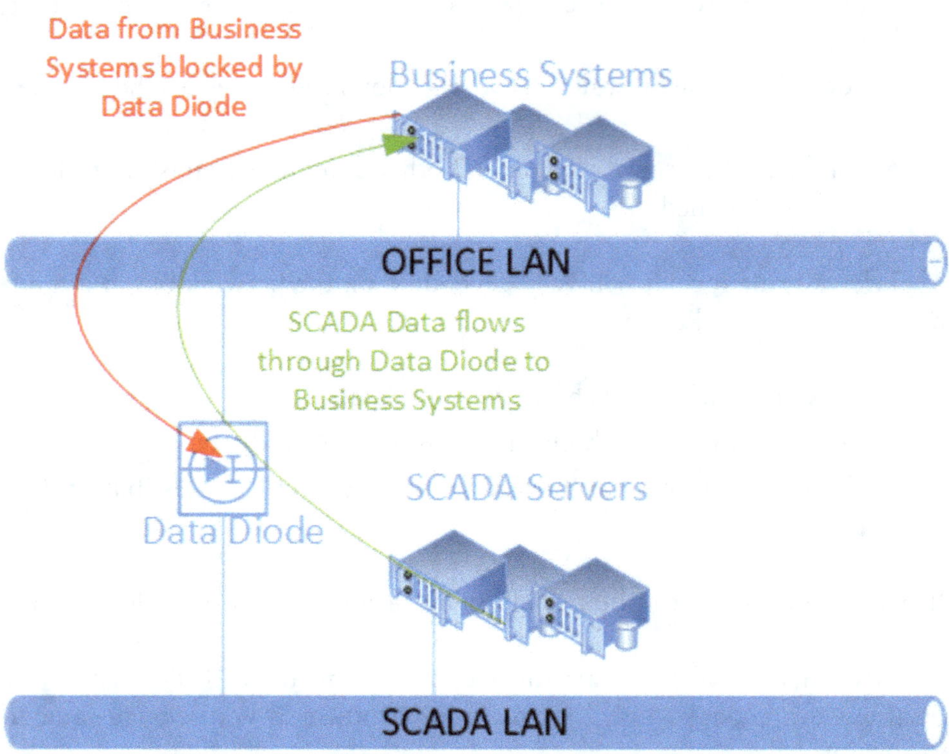

FIGURE 4.2 Data Diode Application Protecting SCADA System

Because the office network is connected to the Internet, if a device on the office network was compromised, then it is much easier for other equipment to be compromised on that network.

For example, CCTV cameras are known to be extremely vulnerable to many types of exploits. This is partly because the hardware and software technology used by many vendors is the same and has the same flaws such as fixed usernames and passwords.

If a CCTV camera is compromised by an attacker by using the default username and password, then that attacker can find other devices, such as the billing and payments workstation, on the network and potentially compromise those too.

To protect against this, it is recommended that systems providing different functionality are separated. In the case of the office network, this would mean that the CCTV system should be separated from the billing and payments system. In addition, it makes sense to separate the office workstation from both as this is used for different purposes from both CCTV and billing and payments.

Separate physical networks can be used to achieve this aim, but this can be expensive to deploy and maintain. For smaller WWUs, the VLAN approach can provide sufficient separation between systems. Virtual LANs are configured in network devices such as firewalls and routers to separate systems electronically on the same physical network.

Note that to support VLANs, it may be necessary to choose a different firewall from the one currently in use. For example, some basic firewalls included in Internet gateways will not include this functionality.

Figure 4.3 shows how the office network could be segregated using VLANs. Note that to further segregate systems, the billing and payments system and the CCTV system that were both previously running on the same office server have been moved to their own servers.

Devices in VLANs may communicate with devices in other VLANs if necessary, and this will be defined in the firewall. As with communication to the Internet, the rules should be very narrowly defined to minimize the possibility of a compromise in one VLAN affecting another.

3.1.1.4 Demilitarized Zone

A "demilitarized zone" (DMZ) is a particular type of network used in secure designs. The name comes from its use in agreements between nations that define boundaries between two or more military powers or alliances. One of the most famous DMZs is the area between North and South Korea.

In computer networks, a DMZ is a network zone that separates trusted and untrusted networks. Demilitarized zones are implemented using firewall(s). A

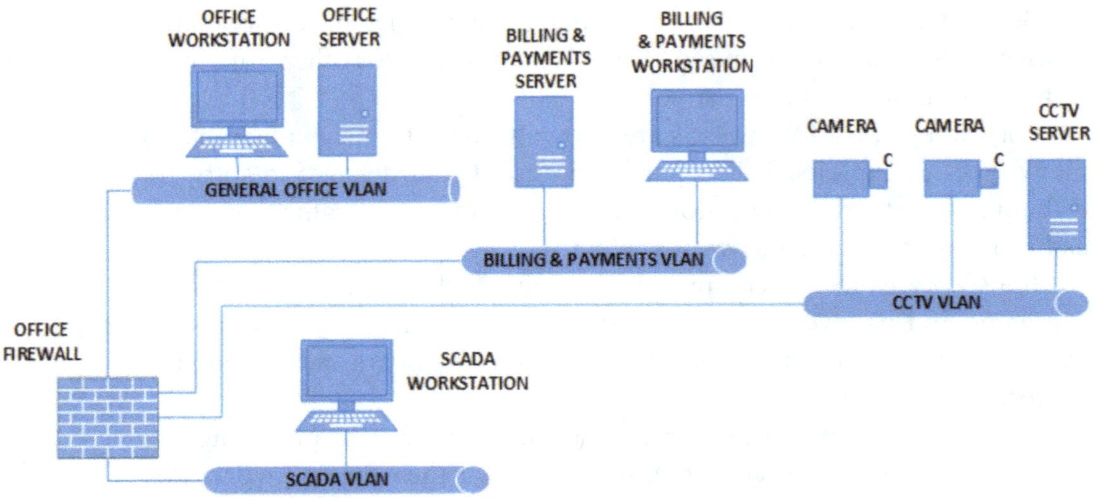

FIGURE 4.3 Example of VLAN Segregation in Office Network

typical example is the separation of a company internal network (the trusted network) from the Internet (the untrusted network). In this scenario, the DMZ provides an additional layer of security: internal resources in the company are not directly connected to the Internet, where they could be more easily accessed. Instead, the internal resources connect to other resources in the DMZ, and these resources in the DMZ connect to the Internet. A simplified block diagram of a DMZ arrangement is shown in Figure 4.4.

3.1.2 *Software Design Considerations*

When implementing or changing systems, there are some key considerations:

- Limit what can be changed. An example is defining limits on setpoints in SCADA systems. It should not be possible for anyone to set a value that creates a hazardous situation (for instance, a toxic level of a chemical).
- Limit who can make changes. An example is limiting who can transfer funds from a bank account. This should not be an action available to anyone who has access to the billing and payment system.
- Separate electronic authorization for actions, such as those above, to change setpoints in a control system or make payments. It may be that there are certain changes that can be made by some users, and certain changes that require secondary authorization.
- Limit what can be done remotely. If there is a valid business need to have remote access to systems, clearly define what is allowed remotely. For example, it may be undesirable to allow changing of setpoints

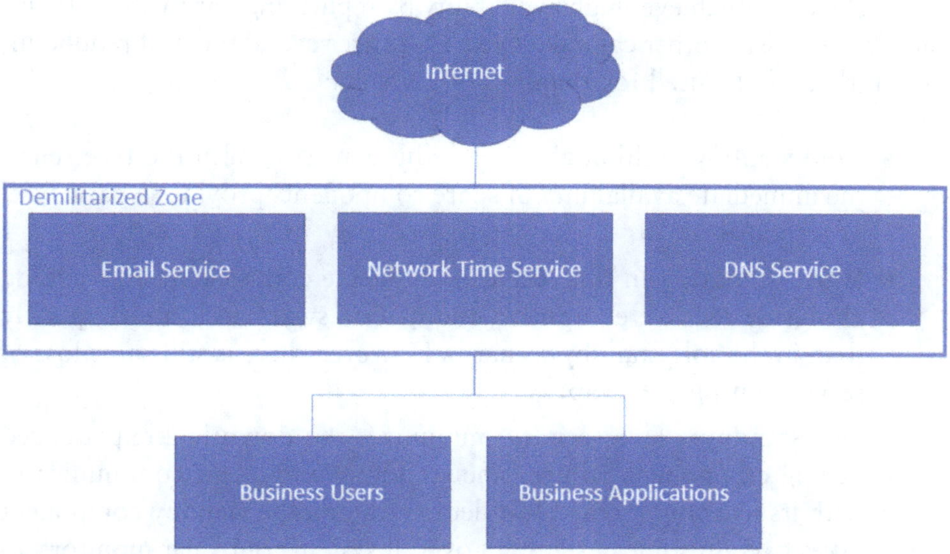

FIGURE 4.4 Demilitarized Zone

remotely because there is a risk that these changes could be made by unauthorized users. There should be different types of remote access users defined if privileged actions are necessary remotely. This is discussed later in this chapter.

3.1.3 Availability, Reliability, Maintainability, and Redundancy

Failure of hardware or software components can result in a major safety or operational issue in a WWU. Cybersecurity threats can cause a failure of components, as can vulnerabilities such as hardware or software flaws. Designing and maintaining systems with availability, reliability, maintainability, and redundancy is critical to achieving safe operations in a WWU.

Availability is the probability that the system is operating properly when it is required. Availability is measured as a percentage over a defined period, for example, per day, per month, or per year. Availability over a year is the most frequently used measure for a system. An availability of 90% translates to 40 days per year downtime. An availability of 99.9999% translates to 30 seconds downtime in the same period. Availability is the combination of *reliability* and *maintainability*.

Reliability is a measure of the probability of a component or system to perform a required function in a specific environment for a specific period without failure.

Maintainability measures the ease with which a product can be maintained and is an essential element for successful operations.

Redundancy achieves high availability by replicating hardware so that if one device fails, another can take over. There are several types of redundant design that can be used for elements of a system:

- Cold standby—Although not a redundant system in the true sense, the immediate availability of spare components provides a basic level of response.
- Warm standby—In this scenario, duplicate components are running alongside the live equipment and can be swapped in more quickly than in the cold standby scenario. However, there is still some loss of service during the swap.
- Hot standby—This scenario minimizes the downtime experienced during component failure. The standby component communicates with its live counterpart. If it detects failure, the standby component takes over. In some designs, an overall system controller monitors all components to detect failures.

Table 4.1 lists the key considerations to maximize equipment availability. The high-level risk assessment performed as part of the cybersecurity risk management process will help identify availability requirements for systems at each location. These requirements can then be provided to vendors and verified during acceptance testing.

3.1.4 Data Security

3.1.4.1 Classification of Data

Water and wastewater utilities will send, receive, and store a variety of data. Each type of data will require different levels of security to protect it. Table 4.2 provides recommended classifications of each type of data in a WWU.

The classifications above are based on the following definitions:

- Public—As the name implies, these data are openly shared (e.g., on websites or through blanket customer emails). These types of data does not require any specific security controls.
- Internal—These data should be contained within the WWU and should be shared outside the system only on a need-to-know basis. This sharing should be approved by an authorized individual in the WWU before sharing. Other than this practice, these data do not require any specific security controls.

TABLE 4.1 Key Considerations to Maximize Equipment Availability

Scope	Considerations
Power	• An uninterruptible power supply (UPS) should be used to provide temporary power to critical equipment. • Larger sites with bigger demands may require stand-alone generators (e.g., diesel) to provide backup power. • Diverse power supply options will avoid a single point of failure. Alternative options, such as solar and wind, might be more suited to remote locations where generators would be cost-prohibitive. • Monitoring of power supplies is essential. A failure of the primary power supply should be detected and reported. • Monitoring of backup supply options (e.g., UPS battery capacity, generator fuel capacity) makes operators aware of remaining capacity.
Communications Networks	• A *stacked* switch combines multiple, individual switches to create a single switch. A chassis-type switch with multiple modules provides improved reliability. In this design, an individual module can be replaced without taking the remaining modules out of service. • Redundant firewalls can be configured to operate in active/passive mode or active/active mode. In active/passive mode, one firewall is active and processes all data while the other is in passive mode. The passive firewall is synchronized, but inactive, unless the primary firewall fails. In active/active mode, the processing load is shared between both firewalls. If one firewall fails, the other automatically takes on the processing for both until the failed device returns to service.
Servers and Workstations	• The SCADA data collection servers should be capable of operating in a hot standby arrangement, with automatic failure detection and switchover. • Servers should be deployed at different geographic locations, if possible. This minimizes the likelihood of total loss of service and allows operation from various locations in the event of a disaster. • Disk storage should be high availability, for example, a redundant array of inexpensive disks (RAID). RAID offers seven levels, 0 to 6, of progressively higher availability. The precise level of redundancy must be balanced against the additional cost, space, and power.

(*continued*)

TABLE 4.1 Key Considerations to Maximize Equipment Availability (*Continued*)

Scope	Considerations
Automation Devices (PLCs, RTUs, etc.)	• Architecture—For most sites, it will be acceptable to have a cold standby arrangement for processor units, input/output (I/O) modules, and communications modules. In certain scenarios in which very high availability is required, warm or hot standby architectures may be mandatory. • Power supply—In hot standby architectures, a power supply shared between the redundant components can represent a single point of failure and should be avoided. Even in non-hot standby architectures, redundant power supplies are recommended as they tend to have a higher failure rate than other devices. • Data logging—Depending on the recovery objectives, it may be necessary to log and store data in the event of equipment failure. For instance, if there is a power failure, nonvolatile data storage will ensure the data up to the power failure are not lost. It is important to understand how the SCADA system deals with logged data. In some systems, data are time-stamped by the PLC/RTU when they are collected. In other systems, the data are time-stamped when they are received by the SCADA system. In a power loss situation, the latter arrangement could result in misstamping of data from the PLC/RTU. • Where communications with a site are operations critical, a secondary communications link will be necessary. This link should be different from, and independent of, the primary communications link. For example, if the primary communication is over a fiber network, the secondary communication could be cellular.
Support Contracts	• Support hours—Many systems need 24/7/365 support; although, depending on the target availability, support may be limited to office hours. • Response time—It is important to specify the skills and knowledge needed to provide support and a time window to get the right person on-site. • Spares—To support the target availability, it may be necessary to deploy spares either on-site or at a nearby vendor location. It is essential that spares are properly maintained and accessible when needed.

TABLE 4.2 Recommended Data Classifications for WWUs

Data	Classification
Customer and employee personally identifiable information such as name, address, Social Security number, and so on.	Confidential
Vendor bank details.	Confidential
Customer water meter readings.	Internal
Water and wastewater treatment data such as chlorine residual measurements, pump hours run, and so on.	Internal
The WWU engineering drawings and specifications	Internal
The WWU operational and financial data.	Internal
Protected Critical Infrastructure Information (PCII) shared with federal government organizations such as Department of Homeland Security and Federal Bureau of Investigation.[2]	Confidential
System usernames and passwords.	Confidential
Press releases and customer service notices.	Public

- Confidential—These data are highly sensitive, as their disclosure could result in a cybersecurity incident or could have another serious financial or legal effect on the WWU if disclosed. These data must be used within the WWU on a need-to-know basis. The data must also be protected when stored or transmitted electronically.

3.1.4.2 Encryption

Encryption protects data when transmitted or stored. Encrypted communications protocols such as HTTPS (Secure Hyper Text Transfer Protocol, the protocol that is used for most web-based communications) are now routinely used for Internet-based services.

Encryption is a technique for transforming data in such a way that it becomes unreadable so that, even if someone can gain access to the data, they won't be able to do anything with it unless they can decrypt it. The data to be encrypted, also called *plaintext* or *cleartext*, is transformed using an encryption *key*, which is a value that is combined with the original data to create the encrypted data, also called *ciphertext*. The same encryption key

[2]Cybersecurity & Infrastructure Security Agency. (n.d.). Protected critical infrastructure information (PCII) program. Retrieved February 11, 2024, from https://www.cisa.gov/resources-tools/programs/protected-critical-infrastructure-information-pcii-program

is used at the receiving end to decrypt the ciphertext and obtain the original cleartext.[3] This is shown in Figure 4.5.

Note that some legacy systems may use HTTP, an insecure version of HTTPS. This should be verified with the vendor and corrected if still in use. This may require an update to the system or a change to the configuration.

Personal computers manufactured since the early 2000s incorporate a security feature called Trusted Platform Module (TPM), which is a hardware-based security feature that is designed to enhance the security of devices like personal computers. A TPM provides a secure area on the device's motherboard for storing cryptographic keys, securing sensitive information, and performing various security-related functions. It is often used to ensure the integrity of the system and protect against various types of attacks.

In addition to TPM, all Windows Professional editions since Windows Vista include a program called *BitLocker*. This is a disk encryption program that encrypts the entire disk, including the operating system and all the files stored on it. BitLocker works in conjunction with TPMs.

When BitLocker is enabled on a personal computer, encryption keys are generated. These are encrypted and protected with the user's or system's PIN (Personal Identification Number) or password. When the computer boots, the user must enter the correct PIN or password to unlock the data on the disk. When data need to be read from the encrypted drive, BitLocker decrypts it on-the-fly using the key, which is temporarily stored in memory during the session. This allows the operating system and applications to access and use the data without any noticeable delays.

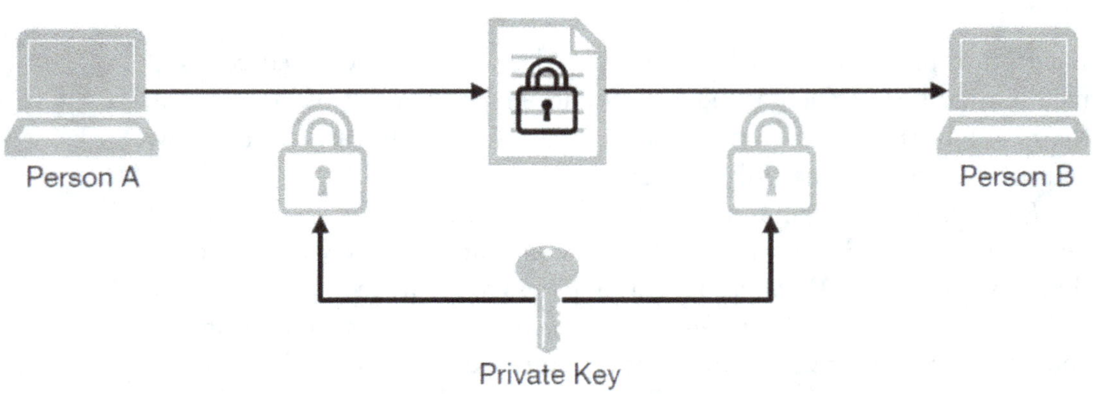

FIGURE 4.5 Encryption

[3]This is a description of *symmetric* encryption. There are more complex forms of encryption, such as *asymmetric* encryption, but the discussion of these details is beyond the scope of this book. Refer to Mustard, S. (2018). *Certified mission critical operations primer* (pp. 64–68). International Society of Automation.

BitLocker continuously protects the data on the disk. If there are any changes to the computer's configuration or the security settings, BitLocker may require the user to re-enter the PIN or provide another form of authentication to ensure that the key remains secure.

Note that it may be necessary to upgrade the hardware and software on machines to use BitLocker. However, this is an essential security control, and the investment is justified.

Encrypting individual files or email messages when transmitting by email can be done using a variety of methods. Microsoft Outlook 365 (when appropriately licensed) supports the encryption of email messages and their attachments. Programs such as Winzip also support file encryption.

3.1.4.3 Hashing

Hashing is a form of encryption. In hashing, an encrypted version of data (the *hash*) is created but cannot be decrypted. An example of hashing is shown in Figure 4.6. Hashing is often used to securely store passwords. In this scenario, to verify a password, a hash is created on-the-fly and compared against the stored hash. This avoids the need to store and transmit the unencrypted version of the password, which could be accessed by unauthorized users.

Another important use of hashing is to verify that files have not been tampered with during storage or transmission. To do this, the file owner

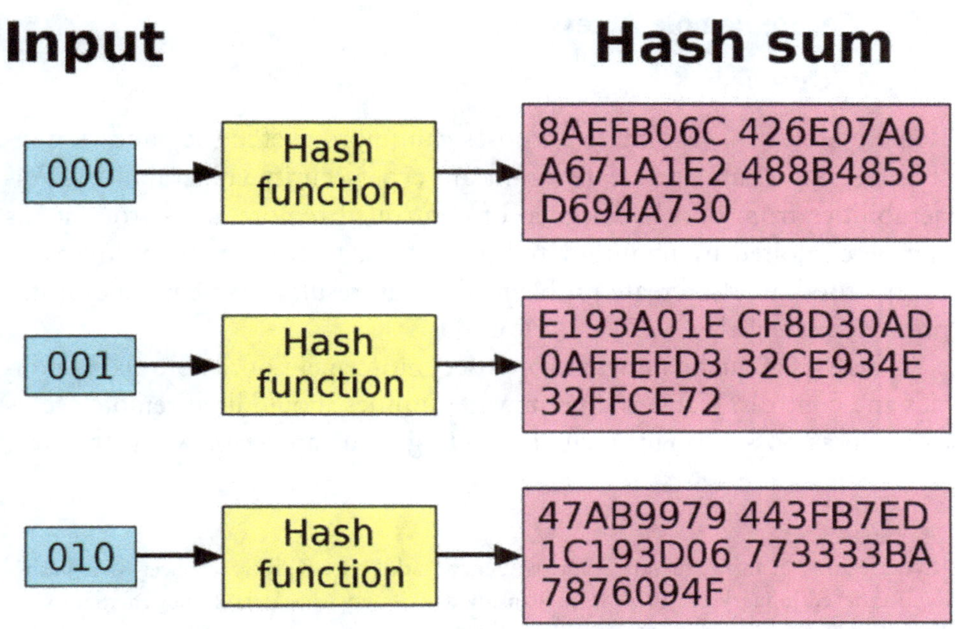

FIGURE 4.6 Hashing

needs to generate a hash of the file using a hashing tool. They then provide the file and its hash. The recipient can run the file through a hashing tool and verify it results in the same hash. If the hash is different, the file owner should be contacted.[4]

3.1.4.4 Recommendations

To protect their data, WWUs should:

- Classify all data using the definitions in Section 3.1.4.1 or similar.
- Protect all data in accordance with their classification when stored and when being transmitted.
- Check all Internet-based communications to verify that they are using secure, encrypted communications.
- Upgrade or replace all applications and services that do not use secure, encrypted communications over the Internet.
- Check all files from vendors by requesting that the vendor provide a hash of the file and verify that the hash of the file you receive matches.

3.1.5 Documentation

Table 4.3 provides a list of documentation that will be required to maintain a secure design for a WWU.

3.2 Secure Remote Access

3.2.1 Overview

Remote access for legitimate purposes can improve efficiency and responsiveness, but it can also create significant cybersecurity vulnerabilities. Vulnerabilities from poorly designed or controlled remote access to systems can be exploited by unauthorized individuals. Authorized employees or contractors can also create problems that can result in system issues if not properly managed.

These risks can be mitigated by strict enforcement of remote access policies and controls. Water and wastewater utilities should limit remote access as much as possible. This includes limiting who can access, what they can

[4]Cybersecurity & Infrastructure Security Agency. (n.d.). NCCIC ICS fact sheet—File hashing. Retrieved January 21, 2024, from https://www.cisa.gov/sites/default/files/FactSheets/NCCIC ICS_Factsheet_File_Hashing_S508C.pdf

TABLE 4.3 List of Secure Design Documentation for a WWU

Document	Purpose
Functional specification	This will document all the technical requirements that were identified and are to be incorporated into the deliverable(s). These will be verified by executing testing using a test specification. This document may include firewall details that are required for secure operation.
Test specification	This document will define the testing to be performed to verify that all the technical requirements in the functional specification have been delivered.
Network diagram	This diagram shows all the key elements of the infrastructure, and the connections between them. See Figure 3.7 in Chapter 3, "Assessing and Managing Cybersecurity Risk."
Physical layout drawings	These drawings will show equipment cabinet layout, building or room layouts.
Wiring diagrams and schedules	These documents will show the physical connection of cables, as well as their locations.
Inventory list	This list will include details of all equipment provided, including manufacturer make and model number, serial number, software, and firmware version numbers, addressing details, and any other salient details, needed to maintain the equipment.
Operations manual	This document will provide details of how the equipment operates, any maintenance that is to be performed, and how to take and restore backups. The Incident Response Plan or Disaster Recovery Plan may refer to this manual or may incorporate elements of it, as needed.

access, when they can access, from where they can access, and what they can do with that access.

Key objectives in providing secure remote access are:

- All non-essential remote access connections to systems are removed.
- All essential connections to systems are documented and approved.
- All essential connections are secured in accordance with standards and recommended practices.
- Systems are segregated using firewalls, VLANs, and other methods described earlier in this chapter.
- Remote access is limited to authorized and competent persons carrying out specific approved tasks.

3.2.2 Types of Remote Access

There are various types of remote access. The most common are

- use of a system application on a laptop, desktop, tablet, or smartphone to perform normal system functions from a location outside the main facility;
- access to a web-based, read-only view of system data, served from a system separate from the main system;
- access to programmable logic controllers (PLCs), remote terminal units (RTUs), or other devices to remotely program or monitor operation; and
- vendor access to systems or devices for equipment maintenance or service purposes.

3.2.3 Key Considerations

Key security considerations for remote access solutions are as follows:

- Eliminate open access to remote equipment (e.g., directly from the Internet). Connections must be made through a secure network connection such as a virtual private network (VPN).
- Incorporate encryption to prevent unauthorized interception of sensitive information, such as device usernames/passwords.
- Adopt *multifactor authentication*. This involves requiring different types of information, such as username/password (something you know) and a secure token or one-time password (something you have).
- Remote access sessions should automatically disconnect after a defined period of inactivity (e.g., 15 minutes).
- Devices supporting remote access should be strictly controlled and secured when not in use. Loss of a device should be reported immediately, and the associated account/device details removed from the system.

3.2.4 Recommended Remote Access Policies and Procedures

The following remote access policies and procedures should be established.

3.2.4.1 User Management

- Each user should have their own user credentials (username/password).

- Remote access should be limited to essential need only.
- Remote access for vendors should be strictly limited to specific, authorized individuals.
- Vendors should not be issued a secure token or one-time password. The WWU should issue a token/one-time password that enables access for each approved work activity.
- Users should be centrally managed so that access may be promptly removed if required.
- There should be automatic logging of remote access (successful and failed logins with addresses, usernames, passwords, time, and duration of session), allowing for review of logs, as required.
- There should be a live view of remote connections accessible from the WWU. This enables connections to be terminated, if required.
- Personnel changes (e.g., transfer, resignation, or termination) should trigger immediate removal of remote access accounts.

3.2.4.2 Approval Process and Oversight

- Remote access connections should be reviewed on a recurring basis. The frequency of review will depend on the risk of compromise of the WWU, and may be monthly, quarterly, or annually.
- Audit logs should be analyzed regularly (monthly or quarterly should be sufficient, unless there are other signs of unusual activity), and issues addressed immediately.

3.2.4.3 Technical Controls

- Remote access should be strictly limited to selected devices via secure VPN.
- Remote access devices should be fit for purpose. Critical features such as screen size and method of data input can affect security and accuracy. A small display and touchscreen, for example, may increase the likelihood of incorrect data entry when paired with read–write access.

In some situations, it may be necessary to apply other controls such as *geofencing*[5] to identify or limit the physical location from where a user can remotely access systems.

[5]Geofencing is the use of location-based services to locate users and use this information to make decisions, in this case to provide remote access. Rouse, M. (n.d.). "*What Is Geo-Fencing (geofencing)?*" WhatIs.com, accessed June 21, 2021.

3.2.4.4 Procedural Controls

- All remote access activities involving changes to SCADA systems, or associated devices (e.g., PLC, RTU), should only be conducted under strict approval and oversight. The approval should identify the planned activities, the associated risks, and any additional controls required.
- No remote access activity should be permitted if the risk of a remote-connection failure would leave the facility unsafe or in an out-of-service state.
- Connections should be enabled or disabled from the remote facility, to ensure that someone within the WWU is aware of the connection being made.

3.2.5 Virtual Private Network

Virtual private networks are used to establish secure communications between geographically diverse networks. This creates a single, *logical* network over external *physical* networks such as the Internet. These secure communications are restricted to an authorized group of users. Common uses for VPNs are

- providing users with a secure method of remotely accessing a system to monitor and/or control systems and
- providing vendors with a secure method of accessing systems or devices for diagnostic purposes.

As shown in Figure 4.7, remote users will authenticate to a VPN server before gaining access to the internal network. The VPN server will be installed either as part of the firewall or as a separate machine. Authorized users will have VPN software deployed on their machines to facilitate this process.

3.3 Network Monitoring

Network monitoring is a broad term that includes

- monitoring networks for problems, such as device failures, heavy traffic, and slow response times; and
- intrusion detection and prevention.

3.3.1 Network Monitoring Tools

Network monitoring tools can provide a wide range of features, including mapping of devices on a network, traffic analysis, network performance,

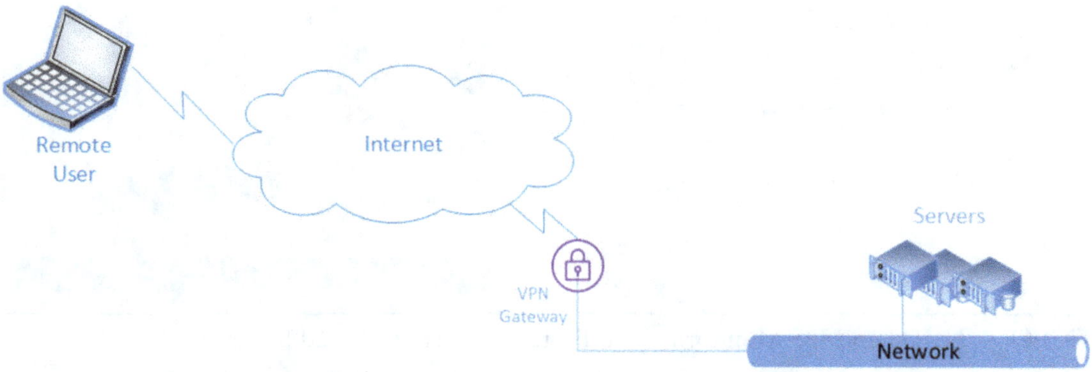

FIGURE 4.7 VPN for Remote Access

user device tracking, and network address management. Firewalls, servers, workstations, and other network-connected equipment produce logs that can be viewed in such tools. These logs help identify suspicious behavior, such as unauthorized attempts to access devices, and unexpected network traffic.

3.3.2 Intrusion Detection and Prevention Systems

An intrusion detection system, or IDS, monitors networks or devices for malicious activity. Networks are monitored by NIDS, or network-based intrusion detection systems, and devices are monitored by HIDS, or host-based intrusion detection systems. An example is shown in Figure 4.8.

An IDS uses signatures, like those used by antivirus software, to detect known attacks. Like antivirus, effective IDS protection requires regular updating of signatures. Also, like antivirus, only known attacks can be detected using these signatures. However, the IDS has a normal baseline for the network or device that it can use to compare current activity against and detect new, unknown attacks. However, in this mode of operation false positives are more likely.

An intrusion prevention system, or IPS, works with an IDS to block known malicious activity when it is detected. An IPS must be used with extreme caution, and only in limited circumstances, in industrial control system environments. Functionality and performance can be severely affected if critical traffic is blocked.

Care must be taken when deploying IDS and IPS technology in an instrumentation and control environment, as follows:

- Some tools can generate significant additional traffic. This traffic can affect the operation of automation system equipment that depends

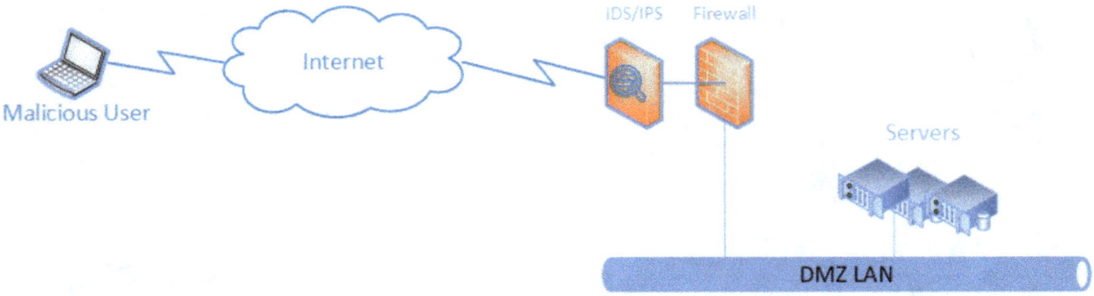

FIGURE 4.8 IDS and IPS Monitoring and Protecting Networks and Systems

on deterministic or near-real-time responses. Passive tools that listen only to data are available to alleviate this issue.

- Often, the operation of automation systems is not well understood. The tools can produce misleading results and false positives. This creates associated inhibition of functionality.
- Procedures are in place to regularly review log files and investigate suspicious activity.

3.4 Physical Access Control

Physical security of facilities is well understood and plays a key role in cybersecurity risk management. Most WWUs have some degree of physical security in place. These may include

- physical barriers or perimeters, such as walls, fences, lighting, and gates;
- security guards;
- locks and electronic ID card access;
- closed-circuit surveillance cameras;
- motion or thermal alarm systems; and
- monitoring storage locations of hazardous chemicals such as chlorine and sodium hydroxide.

Figure 4.9 shows a remote booster station. The SCADA system equipment is housed in a normally locked brick building behind a normally locked fence. Surveillance cameras are monitored and recorded 24/7. This defense-in-depth approach, with multiple controls in place, is highly recommended for the physical security of WWUs' equipment.

FIGURE 4.9 Remote Booster Station

Each control can itself be defeated. For instance, some locks may be easy to pick, chains may be easy to cut, and doors may be easy to force open. Even some types of electronic access cards can be easily cloned with very cheap devices available on the Internet. In the book, *How I Rob Banks (and Other Such Places)*, renowned physical security expert F.C. (short for Freaky Clown, his real name is not made public) describes dozens of real-life experiences testing and defeating physical security controls. The examples highlight the ease with which it is possible to gain unauthorized physical access, even in large and well-funded organizations that invest heavily in security.[6]

Each control should be as secure as it can be (e.g., pick-resistant locks, hardened chains, tough padlocks and shrouds, door hinge jam pins, astragals and dead-latches, and clone-resistant access cards), but the defense-in-depth approach helps provide further layers of protection to make unauthorized access more difficult.

Even in these remote sites, SCADA systems equipment should be secured inside locked cabinets that are inside locked (small) buildings with restricted access. This mitigates inadvertent or accidental cybersecurity issues as well as theft or deliberate attacks.

[6] F.C. (2023). *How I rob banks (and other such places)*. John Wiley & Sons.

Assuming there is some physical security in place, additional considerations for physical security of automation systems equipment include the following:

- Define who should have access to each facility. This might include who has keys, or copies of keys; who is programmed into an electronic card access system; or who has the codes for keypad locks.
- If electronic access is used (in addition to physical keys), electronic access cards should be changed either monthly or quarterly on a regular basis. This may include activities such as changing out cards regularly with new PINs.
- Create a process for taking action when someone leaves. It is common to share codes or keys with employees and vendors. A standard process should ensure that locks or codes are changed, or card access systems are updated, when someone leaves employment.
- Enforce physical security on-site. This includes ensuring that equipment and cabinets are locked when not in use. Visitors must always be escorted to networks from outside the secure perimeter.
- Strict procedures should be in place for bringing equipment to or removing it from a site. Authorized personnel should approve all such activities. The use of unscanned or untrusted laptops and other equipment can result in serious cybersecurity incidents. It may be necessary to prohibit all external devices and use only approved equipment when on-site.
- Have procedures in place for tracking who is accessing a remote site, when, and why. In addition, if a site is accessed outside of expected times, there should be a procedure to respond, possibly involving law enforcement.
- Incident response and disaster recovery plans should consider all cyber, environmental, and physical risks. Exercises should be performed at the facility to ensure the plans are realistic and achievable.

3.5 Electronic Access Control

3.5.1 Overview

Electronic access control enables access to the resources users require, while preventing unauthorized users from gaining access to sensitive information or systems.

The three main elements of electronic access control are *authentication*, *authorization*, and *accounting*.

Authentication identifies a user (e.g., username and password). Following authentication, a user must gain authorization to perform certain tasks. *Authorization* determines which activities, resources, or services a user may access. *Accounting* monitors the resources a user accesses and logs various data. These include system time or data volumes sent and received based on logging of session statistics.

3.5.2 Authentication

3.5.2.1 Passwords

Chapter 2, "The Cybersecurity Risk to Water and Wastewater Utilities," described the many password-related vulnerabilities that exist. Due to these issues, many security experts and organizations are moving toward alternative authentication methods, like biometrics, hardware tokens, and *passwordless* authentication, to enhance security and reduce reliance on traditional passwords. However, passwords are still commonly used, so it is essential for WWUs to do the following:

- Ensure that all manufacturer default passwords are changed.
- Educate users to never reuse passwords and to always use strong passwords; passwords should be at least 16 characters long, as recommended by the Cybersecurity & Infrastructure Security Agency[7] ("longer is stronger") and in the form of a memorable phrase, like *Sunset!Picnic2*. This password contains upper case and lower case letters, numbers, and symbols but is memorable, unlike 5tar&CB%23xyz.
- Protect all password information. It is highly recommended that a password manager tool is used. Pick a well-known password manager that is well-established and has favorable ratings from a variety of trusted sources. Although there is a risk that passwords could be compromised in an attack on the tool provider, the alternatives (such as storing in files or books) carry a far greater risk.
- Implement a Joiners/Movers/Leavers policy that controls who has access to what systems. Any user who has changed roles or has left the organization must have their access removed.[8]

[7]Cybersecurity & Infrastructure Security Agency. (n.d.). *Require strong passwords—Enforcing a password manager protects your business*. Retrieved February 11, 2024, from https://www.cisa.gov/secure-our-world/require-strong-passwords

[8]For further information on Joiners/Movers/Leavers policy requirements, refer to: UK National Cyber Security Centre identity and access management guidance at: *10 Steps to Cyber Security*, National Cyber Security Centre, https://www.ncsc.gov.uk/collection/10-steps/

- Enable multifactor authentication (MFA) on all systems to reduce dependence on passwords alone.
- Provide regular cybersecurity awareness training and perform phishing tests.

Note that the common practice and recommendation for password changes have evolved. Traditionally, organizations and security experts advised users to change their passwords regularly, typically every 90 days. The National Institute of Standards and Technology (NIST) updated its guidelines to move away from frequent password changes. Provided that other password requirements are met (as listed above), then passwords should only be changed if a suspected breach has occurred. The rationale behind these changes is that frequent password changes can lead to predictable patterns, such as users incrementing a number at the end of their password or choosing weaker passwords to make them easier to remember.

3.5.2.2 Multifactor Authentication

Without the added security of MFA, passwords are often the sole barrier to unauthorized access. As noted earlier, there are many potential vulnerabilities relating to passwords, and WWUs often do not have policies to address these vulnerabilities.

Because it is possible to compromise a username and password combination, MFA also requires one or more additional factors. The idea is that an unauthorized person is unlikely to have all the factors. Two-factor authentication (2FA) is the most common. The authentication factors of a 2FA scheme may include

- a physical object in the possession of the user, such as a universal serial bus (USB) drive with a secret token, a bank card, or a key;
- a secret known to the user, such as a username, password, or personal identification number (PIN); or
- a physical characteristic of the user such as a fingerprint, an iris, or a voice.

The most common multifactor method is to use a username and password together with a token generated by an application on a cell phone.

identity-and-access-management, retrieved February 11, 2024; NIST SP 800-53, *Security and Privacy Controls for Information Systems and Organizations* (Rev. 5), https://csrc.nist.gov/pubs/sp/800/53/r5/upd1/final (National Institute of Standards and Technology, December 10, 2020).

The application approach is highly recommended compared to basic text-based MFA (e.g., where a code is texted to a phone) as the latter can be more easily compromised. Multifactor authentication is now widely supported by most vendors and should be a high priority to implement for WWUs.

Fast Identity Online (FIDO) is an open standard designed to enhance electronic access control. Key components of FIDO include the following:

- FIDO U2F (Universal 2nd Factor)—A physical security key for use as a second factor (See Figure 4.10 for an example). Users insert a key into a USB port or tap it to a device to authenticate.
- FIDO2—Allows web applications to interact with FIDO authentication methods. With FIDO2, users can log in to websites and online services using biometrics, PINs, or external security keys without relying on passwords.

Fast Identify Online standards have gained widespread adoption from major technology companies, including Microsoft, Google, Apple, and many others. As a result, FIDO authentication methods are increasingly available for online services and applications, providing WWUs with more secure and convenient ways to protect their systems.

FIGURE 4.10 FIDO-compliant Security Key

3.5.3 Authorization

Authorization controls access to systems or applications. It is a critical aspect of cybersecurity, and another area that creates vulnerabilities in WWUs. After authenticating a user is who they say they are (using MFA to avoid dependence on password only), the next step should be to verify what the user can or cannot access. However, in most cases, this type of authorization is not configured, and anyone with user credentials can access anything. In a very small WWU, it may seem reasonable to have such a simple arrangement. However, as noted in Chapter 2, "The Cybersecurity Risk to Water and Wastewater Utilities," the two most common cybersecurity threats are disgruntled employees or contractors and unintentional acts. It is most likely that a WWU will experience a cybersecurity incident either from

- a disgruntled employee or contractor using their access to misappropriate funds, delete data, or otherwise disable systems, or
- a well-intentioned employee or contractor making a mistake, causing a loss of data or other form of compromise.

An administrator account on a computer running any operating system has full control over that machine. They can create and delete other user accounts, change privileges for other accounts, delete data, start or stop services, and install or uninstall applications. Four of the incidents listed in Chapter 2, "The Cybersecurity Risk to Water and Wastewater Utilities," involve disgruntled former employees or contractors who had administrator-level access to systems.

Effective authorization is an excellent preventative control that can be used in a defense-in-depth arrangement to address these threats. The key elements to effective authorization are as follows:

- Application of the Principle of Least Privilege (PoLP). The principle is that individuals should be granted the minimum level of access necessary to perform their tasks or duties, and no more. The principle is designed to reduce the potential damage that could result from accidental or intentional misuse of privileges.[9]

[9]For further information on PoLP, refer to NIST account management guidance in National Institute of Standards and Technology. (2020, December 10). *Security and Privacy Controls for Information Systems and Organizations* (Rev. 5), NIST 800-53. https://csrc.nist.gov/pubs/sp/800/53/r5/upd1/final

- Centralized authorization in place of local authorization. Windows operating systems include security features to administer local user accounts and control what those accounts can access. Local authorization involves significant effort to administer, as each machine must be separately updated with user accounts, password changes, and access levels. As a result, these tasks are often ignored, increasing the security risk. Centralizing these tasks greatly reduces the burden and reduces the security risks. The only exception to this recommendation is if local authorization is necessary for critical processes that are computer-controlled at a WWU facility. In this case, the WWU should ensure other controls are in place to manage the risk of unauthorized access.
- Strict oversight of system access, including accounting, Joiners/Movers/Leavers policies, and regular audits.
- Regular security training and awareness programs for users to help understand the importance of personal security behaviors.

3.5.4 Accounting

Accounting involves tracking and recording the activities of authenticated users within a system. Accounting helps WWUs maintain accountability, track compliance, and investigate security incidents.

Inadequate accounting may result in a failure to detect suspicious behavior and stop it before a cybersecurity incident occurs. For example:

- logins at unusual times of day or night,
- excessive login failures, and
- creation or deletion of accounts or change of permissions on accounts.

Windows includes a tool called Event Viewer that allows authorized users to view different categories of security-related events. Figure 4.11 shows an example of an Event Viewer display. Reviewing event logs should be part of a WWU's routine maintenance activities. The task may be performed by an employee or contractor and should be overseen by a senior employee. This oversight approach, called *separation of duties*, is described further in Chapter 6, "Dealing With Third Parties," and is essential in activities in which safety or security might be at risk.

A key requirement for effective accounting is to have a trusted time source that all systems share. Without this, there is no guarantee that timestamps will be correct or usable. Options for time source include network

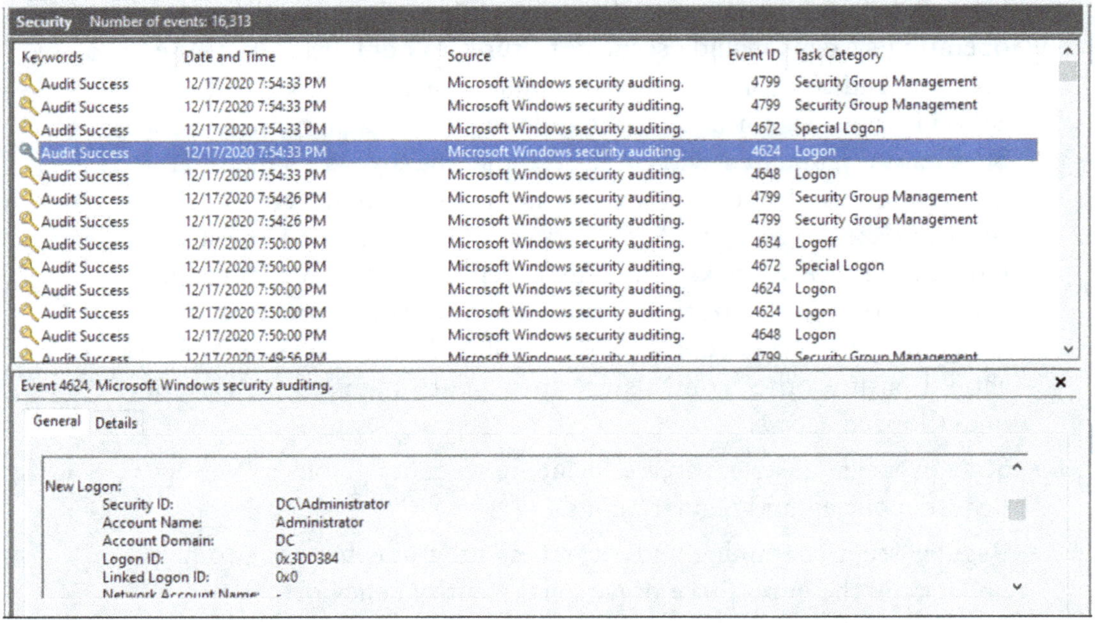

FIGURE 4.11 Windows Event Viewer

time protocol (NTP), Global Positioning System (GPS), or the NIST radio station WWV.

3.5.5 Identity and Access Management

Identity and Access Management (IAM) is a set of technologies and processes that organizations can use to manage and secure access to their systems and data.

Water and wastewater utilities should implement an IAM solution to address the vulnerabilities already discussed in authentication, authorization, and accounting, as follows:

- User Provisioning and Deprovisioning—Centralizing the creation, updating, and disabling of user accounts. A centralized solution would greatly simplify the deactivation or update of user accounts when employees leave or change roles.
- Password Management—The centralized solution would avoid the need to make updates to individual machines. In addition, for users who forget their passwords, self-service options exist allowing users to reset their passwords securely without the intervention of others.
- Implementation of MFA—The use of an IAM solution will automatically incorporate the ability to enforce MFA on all accounts.

- Authorization and Access Control—A centralized IAM solution will allow the configuration of Role-Based Access Control (RBAC), assigning permissions to users based on their roles and responsibilities, or Attribute-Based Access Control (ABAC), assigning access permissions based on attributes of users, such as department, job title, or location.[10]
- Policy Enforcement—The IAM solution will enforce policies, such as password complexity requirements and expiration rules.
- Logging and Monitoring—The IAM solution will capture logs of user activities for security monitoring and audit purposes.

An IAM solution can be deployed on all WWU computers, and even mobile devices. Identity and access management solutions incorporate Mobile Device Management (MDM), which manages access and security settings on mobile devices used by employees.

Deploying an IAM solution may seem expensive or complicated, but modern IAM solutions are cloud-based and relatively easy to deploy. Whereas IAM solutions are not free, the alternative is much more labor intensive and creates far more risks for the WWU.

One of the most well-known IAM solutions is Microsoft's Azure Active Directory (AAD or Azure AD). Because much of the computer equipment in a WWU is Windows-based, AAD is a logical choice. However other non-Microsoft solutions are available, and they can also be deployed in Windows environments. These include Okta, Ping Identity, ForgeRock, IBM Security Identity and Access Management, and Cisco Identity Services Engine. Google Workspace provides its own IAM services.[11]

3.6 System Hardening

3.6.1 General Guidance

System hardening ensures that equipment has all unnecessary programs and services removed or disabled so that the likelihood of using a vulnerable program or service is reduced. The status of equipment should be carefully monitored to ensure that programs are not installed, or services are not enabled, unless they are necessary for the operation of the system and their use has been approved.

---

[10] For further information on RBAC and ABAC, refer to NIST access control guidance in National Institute of Standards and Technology. (December 10, 2020). *Security and Privacy Controls for Information Systems and Organizations* (Rev. 5), NIST 800-53. https://csrc.nist.gov/pubs/sp/800/53/r5/upd1/final.

[11] Inclusion of vendors and their products and services does not imply endorsement.

The following hardening practices are essential for cybersecurity management of automation systems. Certain practices may not be applicable to all devices. The Center for Internet Security (CIS) provides detailed hardening benchmarks for various operating systems, focusing on "technical configuration settings used to maintain and/or increase the security of the addressed technology," and these can be used in conjunction with other controls identified in this chapter.[12]

- Protecting endpoints—This includes antivirus protection and operating system patching, both of which may be performed semiautomatically or manually. In SCADA or billing and payment systems, antivirus and operating system patching should never be performed automatically. Only specific patches and antivirus definitions approved by the vendor should be deployed. Machines should only be rebooted manually. This will avoid unnecessary outage of the system.
- Using USB and network port security—Universal serial bus enabled devices, such as hard drives and devices containing storage, are major sources of malicious programs. A machine should have antivirus or application control installed to prevent malicious programs from executing, but it is recommended that unused USB ports be disabled or locked. USB ports can be disabled in the machine firmware (the BIOS—basic input/output system) or the operating system. They can also be physically locked, using products such as that shown in Figure 4.12. The presence of a lock reminds users to think before plugging their device into a workstation to charge it or download personal files. A lock and key system enables an authorized user to unlock a port if needed. For additional protection, it is recommended that autorun features be disabled in Windows operating systems. This will prevent USB ports, used by mouse and keyboard devices, from intentionally or unintentionally opening USB drives and deploying malicious programs. The same approach is recommended for network ports. Ports on switches and routers should be disabled or configured to *dummy* VLANs that are not used. If someone connects a device to such a port, it cannot communicate without reconfiguration of the network device. Network port locks,[13] like the USB lock shown in Figure 4.12, should be installed to help change behavior and prevent users from connecting without formal approval and change control.

[12]Center for Internet Security. (n.d.). *CIS benchmarks list*. Retrieved February 11, 2024, from https://www.cisecurity.org/cis-benchmarks

[13]For example: https://buy.wesco.com/Automation-and-Control/Motor-Supplies/Interlocks-and-Reversing-Kits/SIEMENS/RJ45-Port-Lock/6GK19011BB500AA0/p/80476625672-1. This is not an endorsement of this product.

FIGURE 4.12 USB Lock

- Using supported operating systems—There are many systems in WWUs today running obsolete and unsupported operating systems (such as Windows XP). These operating systems incorporate known vulnerabilities with no fixes. They may not support the deployment of modern antivirus software to mitigate these risks. Obsolete operating systems may be in service because a system is not supported on newer operating systems, or the upgrade is considered too risky or expensive. The product vendor may not support their product if the underlying operating system is updated. This situation should not be allowed to persist as the risk of operating using obsolete or unsupported operating systems is too great.
- Disabling unused and nonessential features—Windows operating systems come with many applications and services installed and configured by default. Good security practice is to reduce the scope for a cyber incident (malicious or otherwise). Removing programs and disabling services is important, as these are common sources of vulnerabilities. For example, the WannaCry malware exploited a vulnerability in the Server Message Block (SMB) service that runs in Windows XP by default. Many users affected with ransomware were probably

not using the SMB service. Had they disabled it, they would not have been infected. It is not only Windows devices that are affected by unused programs and services. Some PLCs incorporate features such as web servers that are not required. Again, these should be disabled to minimize the attack surface for those devices. Other considerations include removing compact disc or digital video disc drives to minimize the likelihood of introducing infected removable media.

- Configuring resource constraints—Where possible, servers, workstations, RTUs, PLCs, and communications devices should be configured to limit the resources that can be consumed. For example, these devices should be restricted from installing content on a different drive or logical partition other than the operating system. Another option is to place a limit on the amount of drive space that is dedicated for uploads. Ideally, uploads should be placed on a separate drive partition to provide stronger assurance that the drive limit cannot be exceeded. This ensures log files are stored in a location that is sized appropriately. Ideally, log files should be stored on a separate drive or drive partition that is configured to allow a specific number of processes and/or network connections.

System hardening is an ongoing task. It is essential that equipment be monitored to ensure programs are not installed and services are not enabled. Likewise, only approved ports, necessary for the operation of the system, should be enabled. Lack of vigilance is a common security gap in facilities. An end user may put a lot of effort into the initial hardening of systems and equipment. Then someone is inconvenienced by a USB lock. A system service is opened, and the control is never reinstated. Gradually, the hardening erodes until it is ineffective. Figure 4.13 shows an example in which a USB lock has been removed but not reinstated, discovered during a routine audit.

3.6.2 Hardening Wi-Fi

Wi-Fi provides increased flexibility, but presents cybersecurity risks. Key hardening considerations for Wi-Fi include the following:

- Ensure current encryption protocols are used. Wi-Fi encryption standards began with Wired Equivalent Privacy (WEP). This standard has been superseded by Wi-Fi Protected Access (WPA), now in its second generation (WPA2). WEP is considered insecure due to the short encryption key, only 40 bits. This enables hackers to quickly break the encryption. WPA2 uses 152-bit keys. This is so time-consuming to break that they are considered secure.

FIGURE 4.13 USB Lock Removed but Not Reinstated

- The service set identifier (SSID) should not be transmitted. This means that the network is not visible to those users who are scanning for available services. It is still possible to find the SSID, but it can deter casual connection attempts.
- Access points should be configured to accept only devices with specific media access control device hardware addresses. This means that devices must be approved before they can connect.
- Dynamic Host Configuration Protocol (DHCP) that automatically assigns addresses to devices should be disabled. This means a device that is connected is not assigned an address and must rely on a preconfigured, static address. Without an address, the device cannot communicate with other devices on the network.[14]
- Wi-Fi firmware should be kept up to date with the latest security updates. Hardware should be replaced if it cannot support the latest security updates (including changes to encryption standards).

3.7 Supervision and Communications

Cybersecurity, like safety, is an ongoing activity that affects all aspects of a WWU's operations. As a result, an essential barrier is effective supervision and communications. This includes

[14]It is possible to configure DHCP to assign a given IP address to a device based on its Media Access Control (MAC), or hardware, address. However, this can be bypassed by spoofing a MAC address.

- awareness and training for employees and third parties such as vendors, contractors, and suppliers;
- third-party management, ensuring that vendors, contractors, and suppliers have the necessary cybersecurity management practices in place in their organizations; and
- oversight, to ensure that processes and procedures are being followed.

Chapter 5, "Awareness and Culture," provides guidance on how to implement awareness and training, as well as establish a culture of good cybersecurity management.

Chapter 6, "Working With Third Parties," provides guidance on how to ensure that good cybersecurity management practices are in place at vendor, contractor, and supplier organizations.

Oversight involves procedures to routinely check that other processes and procedures are being followed. These should include checking that

- all personnel complete formal training, including ongoing security awareness, and this training is updated annually to keep up with evolving threats, vulnerabilities, and mitigations;
- all site visitors take a cybersecurity induction that covers the key cybersecurity rules;
- all changes follow a formal change-control procedure that includes updating and approving all documentation;
- all changes involve backups of equipment before and after the change;
- protected removable media are secured when not in use and are reformatted and virus-checked before each use; and
- incoming media are scanned for viruses and the contents are transferred to protected removable media before being used.

4.0 CYBERSECURITY MITIGATION BARRIERS

It is essential that WWUs understand that even with cybersecurity prevention barriers in place, there is still a non-zero possibility of a cybersecurity incident occurring. As noted in Chapter 3, the barrier model is also called the *Swiss cheese model* because it is known that holes (vulnerabilities) will be present in the barriers from time to time.

As a result, the following cybersecurity mitigation barriers must be installed, maintained, and operated to reduce the effect of a cybersecurity incident when it does occur, whether it be malicious or unintentional, from outside the WWU or from within:

- cybersecurity incident response plan,
- backup and recovery procedures,
- mechanical fail-safes, and
- manual procedures.

4.1 Cybersecurity Incident Response Plan

When a cybersecurity incident occurs, an incident response plan must be initiated. Incident response plans must cover all the failure scenarios considered in the high-level risk assessment. The incident response plan will define

- recovery objectives;
- roles, responsibilities, and levels of authority;
- communications procedures and contact information;
- locations of emergency equipment and supplies; and
- locations of spares and tools.

There are two key recovery objectives to identify:

1. The recovery time objective (RTO)—Defining how long the function can be out of service
2. The recovery point objective (RPO)—Defining how much data can be lost in the event of a failure

These objectives will dictate what must be in place, in terms of

- service level agreements (SLAs) with vendor(s);
- system design (availability, redundancy, etc.);
- spare parts, on- and off-site;
- what is backed up and how often; and
- how long the backup and restore takes, and the backup location(s).

In extreme cases, system unavailability may become a disaster-level situation, for instance, the loss of a building in a flood or fire. This scenario requires specific recovery actions. A disaster recovery plan should be produced, defining these disaster scenarios and the required actions.

The incident response plan and disaster recovery plan should be exercised on a periodic basis. These exercises verify that the procedures work, and the documentation provides adequate detail. Varying scenarios should be chosen, and the drills should be as realistic as possible. The frequency of drills will depend on the organization. Many critical infrastructure organizations undertake these drills once or twice a year.

Incident response plans and disaster recovery plans should incorporate NIMS, the National Incident Management System. "NIMS guides all levels of government, nongovernmental organizations and the private sector to work together to prevent, protect against, mitigate, respond to and recover from incidents."[15]

4.2 Backup and Recovery Procedures

In the event of a cybersecurity incident, it is essential that systems be restored to known working order as soon as possible. Water and wastewater utilities should never pay ransoms to recover their systems or data. Paying ransoms encourages further ransom attempts, and there is no guarantee that paying a ransom will result in successful system or data recovery.

Backups are, therefore, essential if WWUs are to be able to restore systems or data after a cybersecurity incident.

It is essential that

- backups are taken at a frequency commensurate with changes in the system, so that restoration does not lose a significant number of changes;
- backups are logged, labeled, and stored in a readily accessible location, with copies in alternate locations in the event of a major disaster; and
- backups are tested regularly to ensure that they will work in the event they are required.

Table 4.4 shows the types of backups required in a WWU, and the typical backup type and frequency.

Each equipment type should have its own procedure that describes the specific steps taken to perform the backup.

Backup files must be kept off-site to protect against a localized disaster that could destroy the equipment and backups. Depending on recovery objectives, it may also be necessary to hold copies of backups locally to

[15] Federal Emergency Management Agency. (n.d.). *National incident management system.* Retrieved February 11, 2024, from https://www.fema.gov/emergency-managers/nims

TABLE 4.4 Types of Backups and Their Typical Frequency in a WWU

Equipment Type	Backup Type	Backup Frequency
Server (SCADA, Billing System, etc.) and application software	Disk image	On change (Windows updates, application updates)
Database	Data/file backup	Based on update frequency (e.g., daily/weekly/monthly)
Workstation	Disk image	On change (Windows updates, application updates)
Automation device (RTU, PLC, etc.)	Program/configuration file(s)	On change
Network device (switch, router, firewall, etc.)	Configuration file(s)	On change

allow for rapid response. In this case, backups should be kept in a fireproof safe to protect them from damage.

Backups should be held separate from the system they relate to, preferably offline. If a system or network is compromised, it is important to avoid compromise (or loss) of the backup.

Restore procedures describe the specific steps needed to restore the system from a backup. It is essential that these procedures be tested regularly using real backup files. This not only verifies the backups are working, but it also helps verify that the restoration process works.

There should be a procedure to verify that malware is not present in a backup. Procedures should, therefore, include an anti-malware scan before a backup is taken, and a similar scan when restoration testing is being performed.

When backups are taken by vendors or service providers, the WWU will need to check that the vendor or service provider is taking the backups, testing them, and checking them for malware.

4.3 Mechanical Failsafes

Water and wastewater utilities use various mechanical failsafes and safety measures to prevent or mitigate failures, protect public health, and ensure the reliable operation of these critical infrastructure systems. It is essential that these are properly maintained. In the event of a cybersecurity incident, they may be the only barrier left to prevent a serious incident affecting the public, the facility, or the environment.

- Overflow Protection and Alarms—such as overflow weirs and independent alarms;

- Check Valves—also known as backflow preventers, these are used to ensure that water or wastewater flows in one direction only;
- Pressure Relief Valves—to relieve excess pressure within pipelines and tanks. They prevent over-pressurization, which can lead to equipment failures or ruptures;
- Burst Discs—designed to rupture at a predetermined pressure to release excess pressure safely;
- Independent Pressure Sensors—monitor pressure levels and can trigger actions, such as shutting down pumps, when pressures exceed safe limits;
- Flow Control Valves—regulate the flowrate of water or wastewater in pipelines and prevent pipe bursts due to excessive pressure;
- Isolation Valves—placed within water and wastewater systems to isolate sections of pipelines or equipment in case of leaks or emergencies. They allow for the isolation of problematic areas while maintaining service to other parts of the system;
- Float Switches—used in sewage and water treatment tanks to trigger alarms or equipment shutdowns when liquid levels reach a critical point, preventing overflows or equipment damage;
- Air Relief Valves—used to release trapped air pockets in pipelines, preventing airlocks and ensuring a continuous flow of water; and
- Dams and Levees—in large water reservoirs and wastewater treatment ponds, dams and levees are constructed to contain water and prevent overflows or breaches during heavy rainfall or flooding.

Regular inspections, maintenance, and monitoring are crucial to ensure that these failsafes operate effectively, when needed, to minimize the effect of a cybersecurity incident.

4.4 Manual Procedures

In the event of a serious cybersecurity incident, for instance one that takes the SCADA system out of service, it may be necessary for the WWU to initiate manual procedures. For example:

- Operate a water treatment plant without automatic control.
- Operate a sewage lift station or booster pump station with external pump and power supply equipment.

- Deliver potable water to customers using other means (e.g., bottled water).
- Calibrate and test instrumentation and inspect mechanical failsafes.[16]

The high-level risk assessment should identify the possible failure scenarios that could affect the WWU. There should be manual procedures documented for each scenario, and these should be exercised periodically to check their validity as well as to provide training to employees should they need to act in future.

5.0 SUMMARY

This chapter has provided an overview of the cybersecurity barriers required in a WWU. The cybersecurity barrier model for WWUs is a way to highlight all the barriers, or controls, that must be in place to provide good management of cybersecurity risk.

The barrier model identifies prevention barriers, those that reduce the likelihood of a cybersecurity incident, including secure design, secure remote access, network monitoring, physical access control, electronic access control, system hardening, and supervision and communications. The barrier model also identifies mitigation barriers, those that reduce the effect of a cybersecurity incident when it does occur. These are cybersecurity incident response plan backup and recovery procedures, mechanical failsafes, and manual procedures.

This chapter described each of these barriers in more detail, highlighting best practices and common issues.

The next chapter will discuss awareness and culture, key aspects that are needed if these barriers are to be effective in the WWU.

[16]Cable, M. (2005). *Calibration: A technician's guide*. International Society of Automation.

Chapter 5

Awareness and Culture

1.0 INTRODUCTION

Cybersecurity awareness training is a common tool used by many organizations. What constitutes awareness training and who receives it can vary considerably. Any cybersecurity awareness training is better than none, but training designed for those in information technology environments is not sufficient for those in operational technology environments. This distinction is lost in many organizations in which training, like other aspects of cybersecurity, is managed by the information technology function. Such generic training neglects the operational and cultural differences in operational technology facilities.

This chapter will identify the operational and cultural differences between an information technology and an operational technology environment. Taking these differences into account, it will explore the essential elements of cybersecurity awareness training and monitoring required for an operational technology environment.

2.0 THE IMPORTANCE OF AWARENESS

2.1 Background

Visit any water and wastewater utility (WWU) today and you will likely find several obvious cybersecurity policy violations or bad practices. Typical examples include the following:

- poor physical security, such as unlocked equipment rooms or keys permanently left in equipment cabinet doors;
- uncontrolled removable media (such as universal serial bus [USB]) used to transfer data;
- lack of procedures for removing physical and electronic access for people leaving the organization; and
- poor electronic security, such as leaving user credentials visible. See Figure 2.4 in Chapter 2, "The Cybersecurity Risk to Water and Wastewater Utilities," for a real example of this bad practice.

Most individuals working in WWUs will be familiar with one or more high-profile cybersecurity incidents; some may have been affected by one. Despite this, the types of violations or bad practices referenced above continue to occur. Lack of training and awareness is a major contributor to this problem. Factors include

- lack of awareness of how vulnerabilities occur;
- the assumption that the likelihood of a cybersecurity incident is low because the WWU is not a target, or it has not happened in the past;
- the assumption that cybersecurity incidents only occur because of intentional acts from hackers; and
- underestimating the consequence of a cybersecurity incident.

2.2 Understanding How Vulnerabilities Occur

Figure 5.1 shows that the cybersecurity risks in a WWU are an accumulation of factors introduced throughout the supply chain responsible for delivering hardware and software products that are used to operate the system.

Product vendors as well as system integrators, engineers, and other service providers must have good training and supervision so that they can design products with minimal vulnerabilities or flaws. These organizations must also have good development procedures in place to develop and test

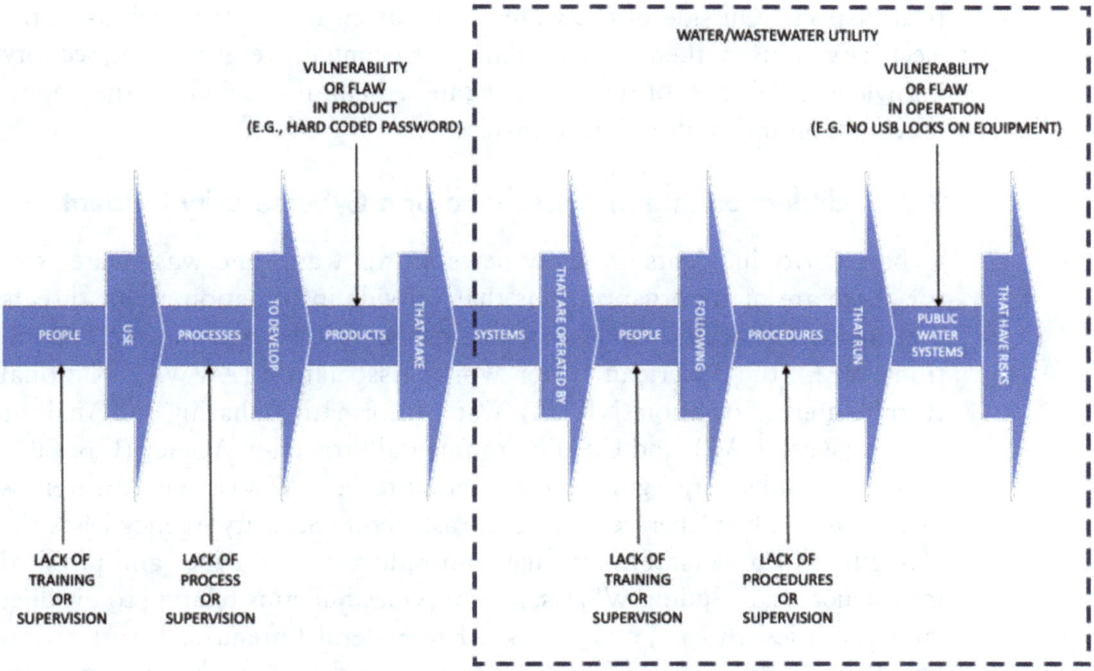

FIGURE 5.1 How Vulnerabilities Occur

products and systems, further ensuring that vulnerabilities do not creep in through the development process.

Although the WWU may have little or no influence in these aspects, employees should be aware that vulnerabilities in products and systems will be present and so they need to take care to avoid these flaws being exploited in a cybersecurity incident.

For example, employees may want to perform additional testing before accepting new products or systems to see if certain common cybersecurity vulnerabilities are present (e.g., is it possible to plug in a USB drive, is it possible to login with a familiar generic username and password).

More importantly, employees who are aware that products and systems have vulnerabilities will understand the importance of the additional cybersecurity barriers or controls that are in place. For example, the USB ports on a workstation may be locked, but the employees understand why and, therefore, ensure they always remain locked.

As shown in Figure 5.1, within the WWU there is a need for training, supervision, and procedures. Gaps in training, awareness, procedures, or supervision can result in cybersecurity incidents. The WWU and its employees can fully control this aspect and so they should focus their efforts on these aspects. Outside of the water and wastewater sector, such as in the electricity industry, there are requirements to annually recertify cybersecurity knowledge. This sort of requirement can be helpful to reinforce the importance of continued attention to this issue.

2.3 Understanding the Likelihood of a Cybersecurity Incident

Cybersecurity incidents are daily news. In the water and wastewater sector, there are many organizations that provide information about threats and incidents specifically in the sector. The Water Environment Federation (WEF), the American Water Works Association (AWWA), National Rural Water Association (NRWA), Water Information Sharing and Analysis Center (WaterISAC), and U.S. Environmental Protection Agency (U.S. EPA) all provide substantial guidance and news to help WWUs with awareness of this issue. The Cybersecurity & Infrastructure Security Agency leads the U.S. effort to understand, manage, and reduce risk to cyber and physical infrastructure, including WWUs, and provides bulletins relating to product vulnerabilities, threats, and attacks. The Federal Bureau of Investigation operates a function called InfraGard to join local alliances between government and the private sector and share cybersecurity threat information for the benefit of the entire nation.

Despite all these resources highlighting the significant threat that exists, WWUs often underestimate the threat to their own organization. The most common reasons for this are described in Table 5.1, with explanations why these reasons are invalid.

The primary reason for these misunderstandings is a lack of awareness training.

2.4 Understanding That Accidental Acts Cause Cybersecurity Incidents

As mentioned in Section 2.3, it is important to consider the entirety of threat sources when managing cybersecurity. Accidental and deliberate acts can have identical consequences. Such accidental acts include the following:

- failure to comply with a procedure, such as those relating to anti-malware, account management, or removable media use;
- failure to maintain or test backups, resulting in an inability to restore a system after a disaster; and
- a device configuration mistake, such as the deployment of an incorrect firewall rule.

TABLE 5.1 Most Common Reasons for Underestimating Cybersecurity Risk

No experience of prior cybersecurity incidents within the organization	• WWUs have limited experience of cybersecurity incidents due to lack of accurate reporting and a failure to correctly identify cybersecurity incidents. • Experience is not a reliable indicator of future results.
A belief that their organization is not a likely target	• Amateur hackers may simply enjoy the challenge of compromising a WWU. • Professional hackers can make money easily through indiscriminate ransomware attacks. • Disgruntled employees or contractors can use their access to get revenge. • Authorized users can make mistakes.
A belief that their organization is secure because of a certain control. For example: "We're protected by a firewall, so nothing inside will happen to us."	• Firewalls protect against unauthorized access from outside the firewall. • Cybersecurity incidents can and do occur within the perimeter of firewalls, for example, accidental introduction of compromised removable media or actions by disgruntled employees and contractors.

As noted elsewhere, accidental acts are a significant contributor to the totality of cybersecurity incidents today and have been since recordkeeping began. Data from various surveys show 75% to 85% of cybersecurity incidents occur because of accidental acts.[1]

To help employees of WWUs understand the significance of accidental acts, it can be helpful to consider the likely causes of various cybersecurity incidents in their environment.

For example, consider the loss of the billing and payments system because of ransomware. Although the ransomware itself may have originated from a professional hacker, it may cause the loss of the billing and payment system through

- an employee clicking on a phishing email and allowing access to the system on their machine,
- a hacker finding an insecure remote access connection to the system or network it resides on, or

[1]Mustard, S. (2022). *Industrial cybersecurity: Case studies and best practices* (pp. 186–187). International Society of Automation.

- an employee failing to scan removable media before plugging into the server hosting the billing and payment system.

2.5 Understanding the Consequences of a Cybersecurity Incident

As explained in Chapter 2, "The Cybersecurity Risk to Water and Wastewater Utilities," the consequences of a cybersecurity incident can be significant. This is true of all asset-intensive sectors, such as oil and gas, energy, and manufacturing. In addition to the typical information technology-related consequences, such as loss of customer or employee personally identifiable information (PII), or loss of functionality such as billing and payments, WWU cybersecurity incidents can lead to the following:

- Loss of potable water supply or sanitary services affecting the local community.
- Compromise of the potable water supply, resulting in harm to customers.
- Damage to infrastructure, such as water mains.
- Damage to machinery, such as pumps.
- Harm to the environment, such as a release of untreated wastewater or toxic chemicals.
- Harm to employees, such as exposure to chlorine gas or another toxic chemical.

It is critical that WWU employees understand the severity of the potential consequences, so they take cybersecurity management seriously.

2.6 Helping People to Visualize Cybersecurity Risk

An important aspect of cybersecurity awareness is finding ways to help people see the connection between their actions and the potential consequences. The bowtie diagram is a useful tool for this purpose. The bowtie diagram is like the barrier model described in Chapter 3, "Assessing and Managing Cybersecurity Risk," and Chapter 4, "Cybersecurity Barriers." The bowtie diagram consists of five key elements, described in Table 5.2.[2]

Figure 5.2 shows a bowtie diagram highlighting the key factors for WWU cybersecurity management.

[2]Bowtie diagrams may include other elements. For this book, the diagram has been simplified to show key elements important to WWUs. Other examples of bowties are available online, for instance: Culwick, M. D., Endlich, Y., & Prineas, S. N. (2020). *The Bowtie diagram: A simple tool for analysis and planning in anesthesia*. National Library of Medicine. Retrieved February 18, 2024, from https://www.ncbi.nlm.nih.gov/pmc/articles/PMC7752234/

TABLE 5.2 Bowtie Diagram Elements and Definitions

Unauthorized access to to Public Water System building	**Threat** This is the source of the cybersecurity incident. It is important to identify each threat as the consequences and the controls may vary.
Public Water System compromised	**Event** This is the situation that the threat creates. There can be many events, but to keep the diagram simple, a single event is identified.
Loss of visibility on SCADA system	**Consequence** This is the negative effect to the WWU from the event occurring. There can be multiple potential consequences for any event.
Incident Response Plan ensures Public Water System can recover in a timely manner with minimal disruption	**Barrier** This is a control that is in place. As described in Chapter 4, "Cybersecurity Barriers," there are prevention and mitigation barriers. As with the barrier diagram, prevention barriers are on the left of the event in the diagram and mitigation barriers are on the right. Prevention barriers reduce the likelihood of an event, and mitigation barriers reduce the effect of the event when it occurs.
Configuration error in IDS, or inattention to alerts	**Escalation Factor** This is a vulnerability or flaw in the barrier. It can be a technical vulnerability (error in software or hardware) or a procedural failure.

The following three threats are considered:

- Unauthorized access to the network from outside the WWU. This is the threat most familiar to those concerned with cybersecurity. It could be an amateur or professional hacker gaining access through various means, including use of phishing emails to employees, or through insecure network equipment. Unauthorized users can also include former employees or contractors who should have had their system access removed when they left the WWU.
- Authorized access to the network from inside the WWU. This is the threat most often overlooked. Authorized users in this context include current employees or contractors. Although authorized, they perhaps

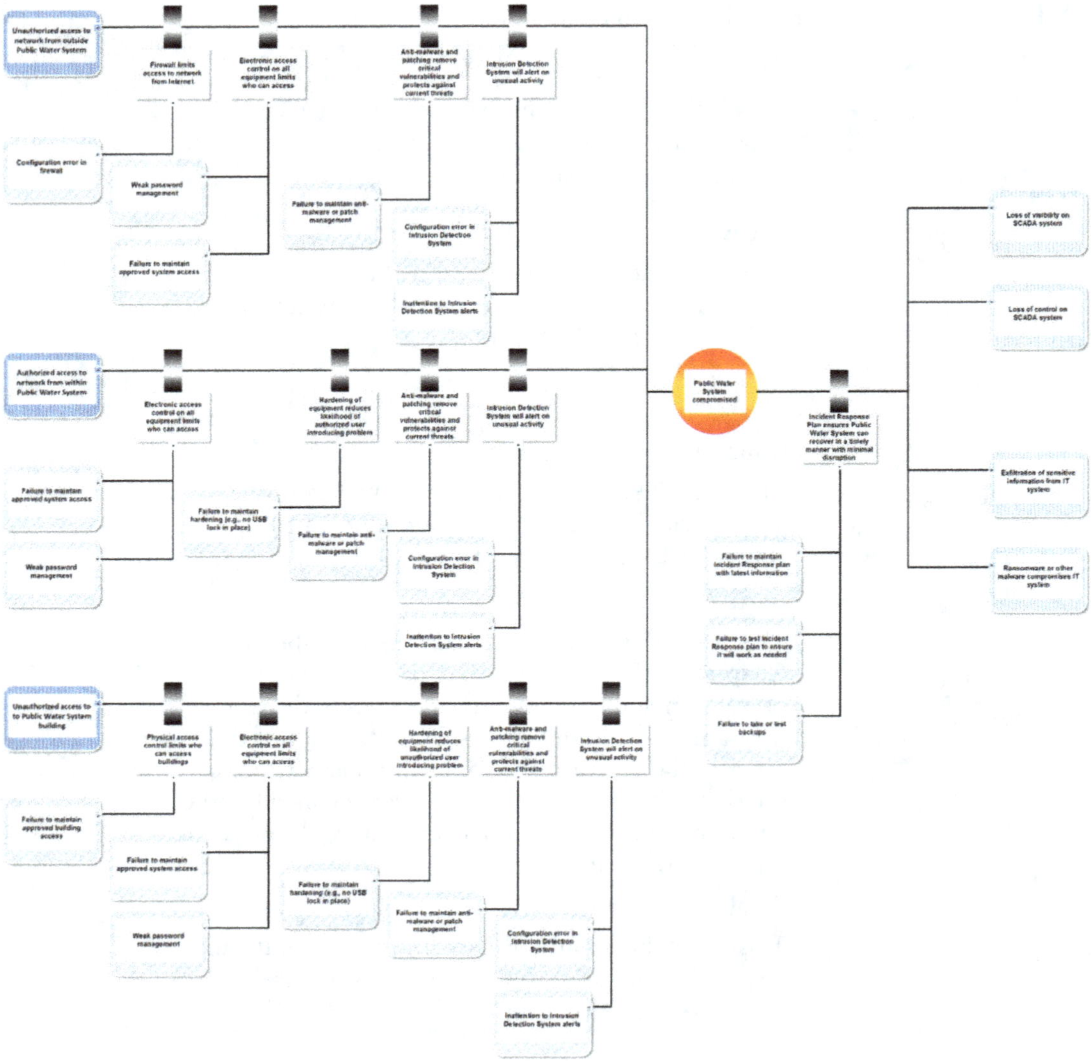

FIGURE 5.2 WWU Cybersecurity Bowtie Diagram

should not have access to certain systems, or they should not perform insecure actions on them (e.g., use a workstation to print personal material brought from home on a removable USB drive). These authorized users can *accidentally* cause a cybersecurity incident.

- Unauthorized access to WWU building. This threat is also often overlooked. Physical security is a key part of cybersecurity management. Unauthorized users in this context include former employees or contractors who should have had their building access removed when they left the WWU.

Four potential consequences are shown:

- Loss of visibility on supervisory control and data acquisition (SCADA) system. This could be caused by ransomware or other malware on all or part of the SCADA system. The result means that the WWU is unable to see the status of processes. Typically, this situation will require deployment of personnel to affected locations to monitor and control processes locally.
- Loss of control on the SCADA system. This would be caused by similar circumstances to loss of visibility. In this case, it may be possible to see the status of processes, but it is not possible to adjust settings remotely. As with the above situation, it will likely require deployment of personnel to affected locations to monitor and control processes locally.
- Exfiltration of sensitive information from the system. This would usually be the result of unauthorized access to a system, possibly through a phishing email attack. The attacker could take sensitive information such as employee or customer PII (name, address, date of birth, Social Security number, etc.), which could then be used to steal identities. Such a situation may require the payment of fines and restitution for those affected;
- Ransomware or other malware compromises system. Compromise of information technology systems such as billing and payment servers or operational technology systems such as SCADA, Computerized Maintenance Management System, or Geospatial Information System can cause severe operational disruption. For example, the WWU may be unable to pay employees or take customer payments. A compromise of such systems may also involve a loss of key historical data that are required to operate, such as previous customer payments. The result of such a situation may involve payment of a ransom to recover data (not recommended) or some period of manual operation while services are restored from backup (recommended).

The bowtie diagram is helpful for awareness as it shows the controls that must be in place to reduce the likelihood of the threats causing an event, and reducing the severity of the event if it occurs. More importantly, the diagram also shows the typical vulnerabilities in the barriers, and how these are mostly a result of human failure or error.

Consider one scenario. Figure 5.3 shows the unauthorized access to the network from outside the WWU threat and its associated prevention barriers.

There are multiple prevention barriers to reduce the likelihood of this threat causing an event, and several involve technical controls that many may feel "manage themselves." However, as Table 5.3 shows, there are

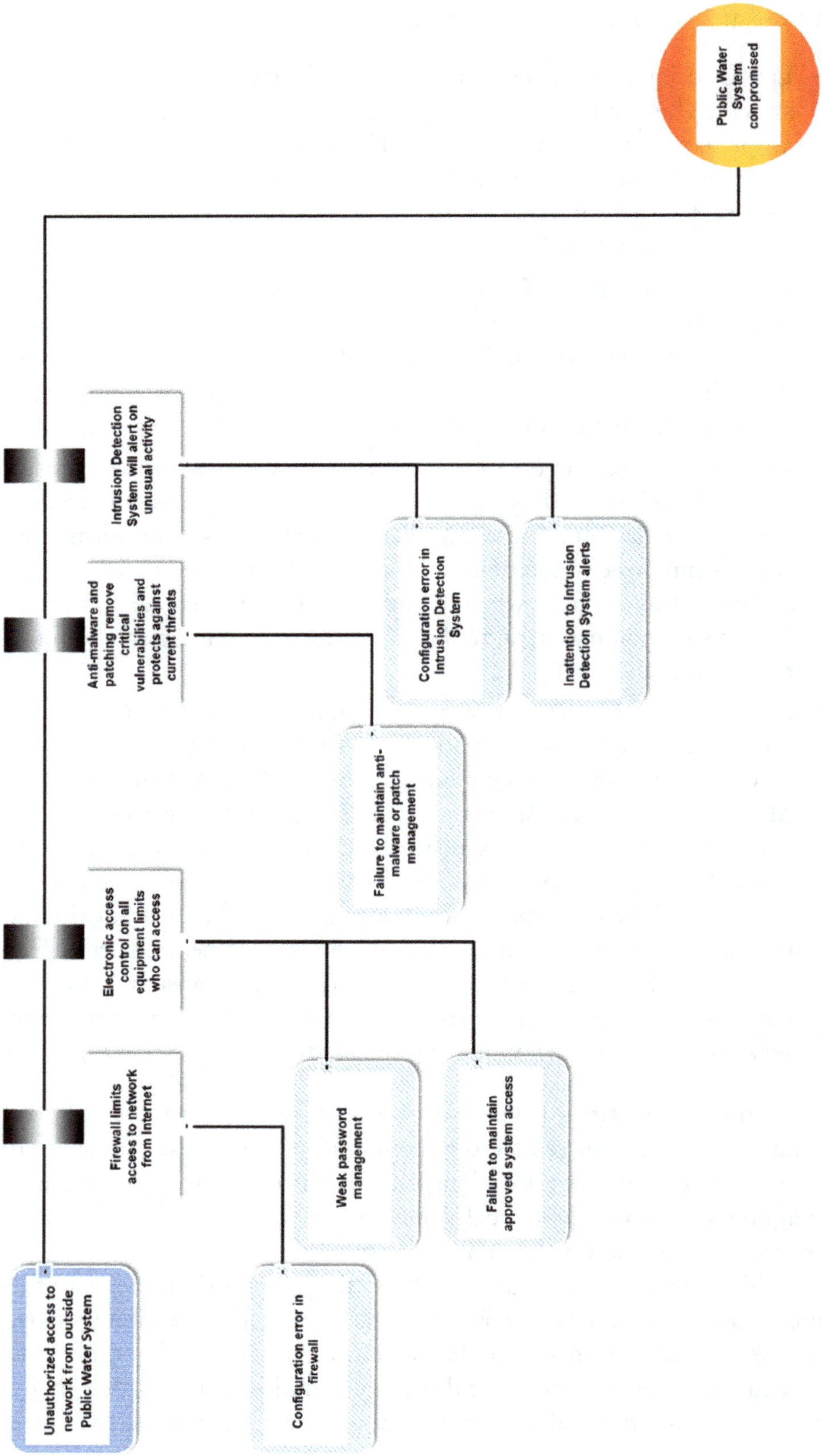

FIGURE 5.3 Prevention Barriers for Unauthorized Access Threat

TABLE 5.3 Possible Escalation Factors Related to Prevention Barriers

Prevention Barrier	Escalation Factor(s)
Firewall limits access to network from Internet	• Configuration error in firewall—firewalls must be programmed with specific rules allowing only authorized users and services. Mistakes in rules can allow unauthorized users to gain access.
Electronic access control on all equipment limits: who can access	• Weak password management—sharing of usernames and passwords, publishing details in the open, failing to change default passwords, and failure to change periodically are examples of weak password management. • Failure to maintain approved system access—access should be limited to only those who need it to perform their job. Users who leave the WWU should have their access removed immediately.
Anti-malware and patching remove critical vulnerabilities and protects against current threats	• Failure to maintain anti-malware or patch management—anti-malware software needs to be updated regularly with new threat signatures. The software itself needs updating periodically, as does computer operating systems and application software such as billing, SCADA, and programmable logic controller programming. Without these updates, systems are exposed to the possibility of compromise through known vulnerabilities in the software.
Intrusion Detection System will alert on unusual activity	• Configuration error in Intrusion Detection System—as with firewalls, these systems are only as effective as their configuration, which must be carefully tested and maintained. • Inattention to Intrusion Detection System alerts—as with burglar alarms, a failure to acknowledge an alert from such a system can mean that a cybersecurity incident can be overlooked until it is too late to prevent.

several possible escalation factors for these barriers that result in vulnerabilities or gaps that can cause the barrier to be ineffective.

All the escalation factors identified above relate to people following procedures, and so the effectiveness of the cybersecurity barriers is entirely dependent on employees understanding their role in protecting the WWU.

This is the same with mitigation barriers, as shown in Figure 5.4. The single mitigation barrier covering all three threats is the existence of an Incident Response Plan. However, several escalation factors for this barrier are commonly observed in WWUs today, as follows:

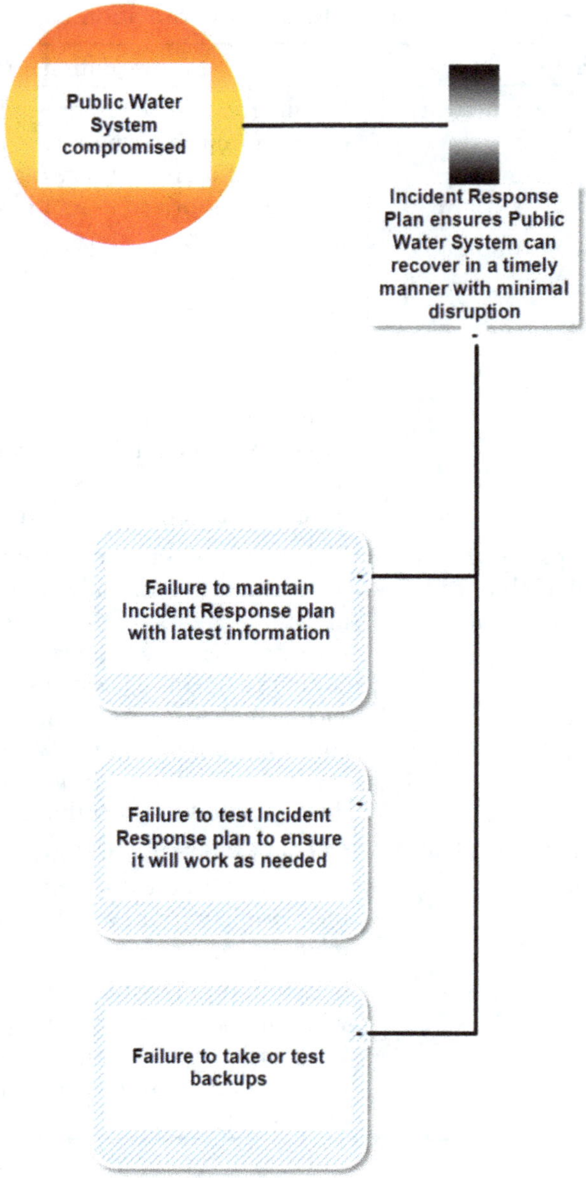

FIGURE 5.4 Mitigation Barriers for All Threats

- Failure to maintain Incident Response Plan. Assuming the WWU has such a plan (and many do not), a failure to update names and contact information when people move on, either in the WWU, local government, or with vendors can result in a response that cannot be executed. A failure to update the plan to take account of new scenarios or results from prior tests can also cause these issues.

- Failure to test Incident Response Plan. Even with a well-written Incident Response Plan, a failure to perform regular and realistic tests can mean that the plan is ineffective when needed. For example, it may be discovered during testing that the only person who knows the password to a key piece of equipment is a vendor. This should result in some changes to ensure that the WWU is not dependent on this individual when there is a real incident.
- Failure to take or test backups. One key aspect of Incident Response Plans is being able to restore systems from backups. The ability to do this depends on whether backups are taken at the correct frequency, stored in accessible locations, and include all elements that may be needed in the event of a cybersecurity incident. Incident Response Plan testing can help reveal gaps in backup strategies.

The use of the bowtie diagram should help WWU employees see that there is a significant dependency on people following procedures. A failure to execute the appropriate actions can result in a cybersecurity incident that leads to potentially severe consequences.

3.0 INSTILLING THE RIGHT CULTURE

3.1 Supporting the Right Behaviors

A key aspect of cybersecurity awareness is helping employees understand their role. Another key aspect is supporting the right behaviors. This support should include the following:

- Providing the tools necessary to maintain security, such as password management tools, secure file transfer facilities[3], and secure, but practical, authentication methods.
- Providing the training required to understand cybersecurity management, and the important role training plays in its success.
- Providing the time needed to maintain cybersecurity controls. Most WWU employees are already overloaded, and so consideration must be given to cybersecurity-related responsibilities. As noted from the bowtie diagram, significant time is required to maintain all the

[3]A secure file transfer facility could be a secure FTP solution with authorization and access control or a solution that allows for the scanning of removable media and the transfer of files from the media into the network in a controlled manner.

necessary cybersecurity barriers such as patch management, anti-malware control, and access control.

3.2 The Safety Culture

Safety culture is helpful in cybersecurity awareness training. Underestimation of cybersecurity risk and overconfidence in the barriers can lead to a lack of concern, or an assumption that "someone else is taking care of those issues." Left unchecked, this lack of concern can be expressed in the acceptance of bad practices such as use of unapproved removable media, leaving cabinets unlocked, leaving controllers open to remote programming, and poor account management.

Just as WWU leaders should "hold people accountable for accomplishing proactive process activities that can prevent injuries," they should apply this equally to cybersecurity activities. Encouraging the reporting of cybersecurity near misses, such as leaving cabinets unlocked, will help to improve the process, and create a more secure culture.

Leaders should allow for *self-directed* behavior, rather than simple compliance with edicts. In the latter case, employees may comply if they are being monitored, but in the absence of management, their behavior changes. For example, an employee transfers files at home between a personal computer and a USB drive without adequate antivirus scanning. If that employee then uses the USB drive at work, it could transfer malware to the WWU.

The more cybersecurity is embedded into the existing safety culture, the more likely it is to be adopted as an integral part of operations rather than an afterthought.

4.0 TRAINING AND AWARENESS

4.1 Overview

Water and wastewater utilities should approach cybersecurity training using a combination of freely available material from trusted sources such as WEF, AWWA, NRWA, WaterISAC, and U.S. EPA and custom material specific to their own organization. This latter material should include specifics on the systems being managed, the specific risks identified, the processes and procedures necessary for effective management of risk, and the Incident Response Plan.

4.2 Who Should Be Trained

Everyone in the WWU should have cybersecurity training. The level of training will vary depending on the role. Table 5.4 shows an example of

TABLE 5.4 Types of Training for Varying WWU Roles

Role	Training
Board members	• Cybersecurity risk to the WWU • Examples • Phishing awareness • Reporting possible incidents or issues • WWU cybersecurity rules
Managers	• Board member training • Overview of procedures for managing security in office (physical and electronic controls) • Overview of procedures run by operators
Operators	• Board member training • Procedures for managing security in facilities (physical and electronic controls)
Office staff	• Board member training • Procedures for managing security in office (physical and electronic controls)
Third parties (vendors, engineers, system integrators, contractors, suppliers)	• WWU cybersecurity rules

varying types of training for varying roles. Note that third parties are also included. They may receive training from their own organization. However, they should receive training on specific rules relating to cybersecurity that they must follow when working on WWU sites or systems.

4.3 Continuous Education and Evaluation

Training is not a one-time exercise. Employees need continuous learning to ensure they are aware of changes to policies and procedures, risks, and barriers. In addition, there must be a system of monitoring to ensure that this training is effective.

A method for tracking cybersecurity effectiveness based on the well-known *safety triangle* is outlined in *Industrial Cybersecurity: Case Studies and Best Practices*.[4]

The safety triangle and this *security triangle* highlight *leading* and *lagging* indicators. Lagging indicators record incidents (of varying degrees of

[4]Mustard, S. (2022). *Industrial cybersecurity: Case studies and best practices* (pp. 43–47). International Society of Automation.

severity) that have already occurred. Leading indicators attempt to show the potential for these incidents to occur before they occur. The leading indicators identified here are as follows:

- Security awareness training, including number of employees who have completed the latest training and performance in various tests.
- Security near misses, which requires the reporting of situations that could have resulted in a cybersecurity incident. For example, use of a USB drive without scanning for malware or leaving a cabinet unlocked.

Changes in these indicators can highlight changes in behavior, such as people being more casual with cybersecurity controls resulting in more near misses. The aim is to identify these changes before an incident occurs, giving the WWU time to correct the underlying behavior.

Two periodic cybersecurity tests that WWUs can conduct to verify behavior are as follows:

- Clear workspace—Periodically checking employee work areas helps to gauge how well employees are following good cybersecurity practices. Common failings include leaving computers unlocked when not

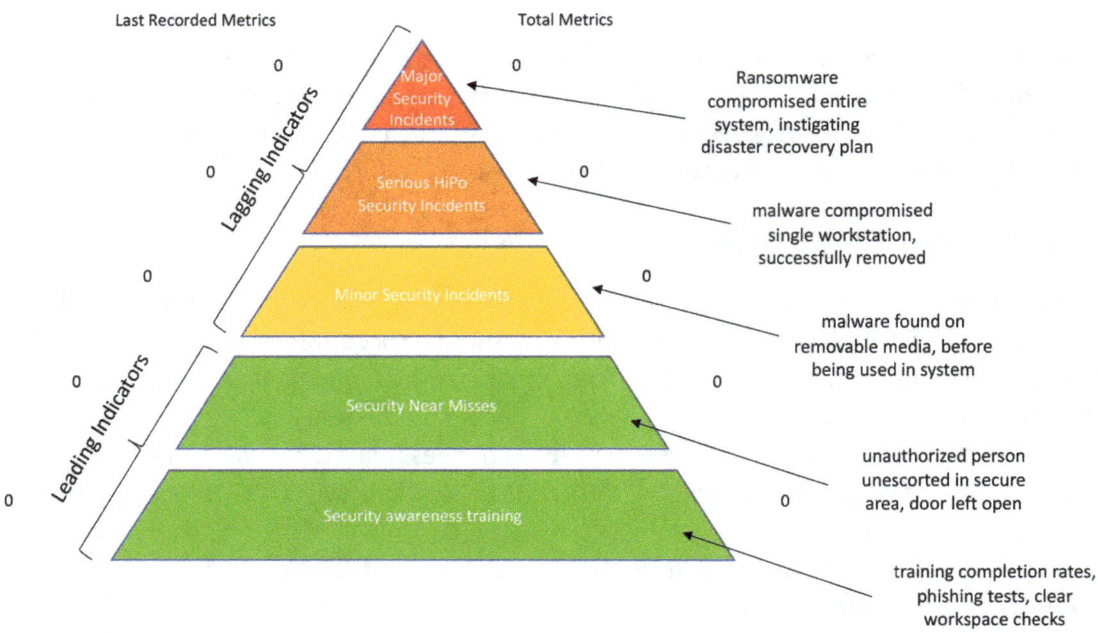

FIGURE 5.5 Security Triangle with Lagging and Leading Indicators

present, leaving sensitive documents on display, and leaving devices such as smartphones and USB drives unattended. Attention to these details helps to drive an improved security culture; and

- Phishing tests—There are a variety of third-party services available to send employees simulated phishing emails, such as fake LinkedIn requests, invoice approval requests, or requests to enter credentials. Although employees can become familiar with this process, it still provides a useful measure of how well people are recognizing social engineering attacks.

5.0 SUMMARY

Cybersecurity is constantly in the news, so it may seem reasonable to believe that WWU employees have a good awareness of the cybersecurity risks their organizations face. However, evidence indicates otherwise as incidents continue to occur. This is largely due to a lack of procedures, employees not following procedures, and a lack of oversight and enforcement by leadership.

A major factor underpinning this problem is that WWU employees underestimate their organization's cybersecurity risk. Another factor is a failure to recognize their role in managing cybersecurity.

Water and wastewater utilities must offer the necessary time and resources for employees to perform their roles. They must also provide the right tools to allow them to do this securely and safely, while still being efficient.

Instilling cybersecurity behaviors in the same way as safety behaviors can be very effective. Employees of the WWU should feel like cybersecurity is part of their job, not an optional extra task to do if time permits.

All those involved in WWUs, from board members, leaders, operators, office personnel, and third parties need training to be aware of the following:

- why cybersecurity is so important;
- how a cybersecurity incident can lead to a serious safety or operational issue;
- what cybersecurity controls are in place, and what happens if they fail; and
- what part each person plays in maintaining good cybersecurity in the organization.

Along with awareness training, it is essential that WWUs identify and monitor leading and lagging indicators that can be used to adjust behaviors before a serious cybersecurity incident occurs.

Chapter 6

Dealing With Third Parties

1.0 INTRODUCTION

Water and wastewater utilities (WWUs) rely heavily on third parties to enable their operations, as follows:

- Engineering contractors design, build, and, in some cases, operate facilities.
- Product vendors support systems and software used in the WWU.
- Service providers provide various functions, including information technology infrastructure or telecommunications equipment support.
- Suppliers provide chemicals and other goods consumed by the WWU.

It is imperative that cybersecurity risk management takes these third parties into account. This involves ensuring that

- contracts are in place, including clauses that set out the required obligations from the third party with respect to cybersecurity risk management;

- requirements for secure design and ongoing secure maintenance of systems and software are spelled out and executed;
- awareness training includes third parties, so that they are aware of the specific cybersecurity risks identified in the WWU and how they can contribute to managing that risk; and
- oversight of third parties is in place to ensure compliance with contractual obligations, policies, and procedures.

2.0 CONTRACTS

2.1 Overview

Third parties may not have adequate cybersecurity management systems in place in their organization. Their vulnerability to cybersecurity incidents exposes the WWU. Key risks to WWUs from inadequate cybersecurity management by third parties are as follows:

- Insufficient controls in place to manage user credentials for the WWU's systems or software. Examples include having standard administrator accounts for all client systems, sharing account details, and not securely protecting these account details.
- Lack of procedures in place to manage or verify backups for WWUs. This could include anything from backups of billing databases to backups of programmable logic controller (PLC) programs.
- Inadequate security awareness training to their personnel. This could result in a third-party employee clicking on a phishing email, eventually leading to a compromise of the WWU.
- Lack of security controls on third-party employee equipment. For example, no anti-malware protection, unpatched operating system software, or inadequate electronic access control.

To manage these risks, there are several minimum contractual requirements that must be in place in all third-party contracts. In addition, depending on the scope of work, there may be a need for additional contractual requirements.

2.2 Minimum Contractual Requirements

The following key aspects must be included in contracts with third parties:

- Provide a named individual with overall responsibility for cybersecurity and authority to escalate security issues or incidents;

- Have a cybersecurity policy developed in accordance with current industry standards (e.g., National Institute of Standards and Technology Cybersecurity Framework, ISO/IEC 27001/2, ISA/IEC 62443);
- Perform good access control practices, as follows:
  - Restrict physical and logical access to sensitive information and systems based on levels of access and privileges required to perform a function or role;
  - Immediately revoke all access for personnel no longer working on the services or those who no longer require access;
  - Review user accounts and privileges on a regular basis to verify that access is correct and remove access that is no longer required;
  - Enforce the use of strong passwords and protect passwords from unauthorized access and interception;
  - Restrict the use of privileged accounts to authorized individuals;
- Conduct a background check in the form of an Identity History Check, including Federal Bureau of Investigation History Check and State History Check and professional reference check, on all their personnel (including its contractors);
- Provide cybersecurity awareness training to all personnel;
- Protect sensitive information in storage and in transit using encryption. Protect sensitive information at every stage of its life cycle from creation through processing, storage, and disposal;
- Manage end-user devices and servers including anti-malware software and patching procedures;
- Maintain security incident response plans to manage the response to incidents and test these plans regularly;
- Report any confirmed security incidents or data breaches to the WWU promptly and without delay; and
- Allow the WWU to perform an audit of the third party to ensure it is enforcing these cybersecurity practices.

These are minimum requirements and should be non-negotiable. Any third party who is not prepared to accept these requirements in a contract should be considered a security risk and the WWU should avoid engaging with them. There may be situations in which the third party does not initially meet the above requirements, in which case the WWU can establish a process to ensure they are met within a reasonable time scale.

2.3 Third-Party Scope

Depending on the scope of work that the third party is delivering, it may be necessary to consider additional contractual controls. Contracts should be tailored to the specific scope that the third party is providing and should be developed using the approach outlined in Table 6.1.

3.0 REQUIREMENTS

3.1 Technical Requirements

Chapter 4, "Cybersecurity Barriers," describes the cybersecurity prevention barriers that must be installed, maintained, and operated. When third-party

TABLE 6.1 Cybersecurity Risks and Controls for Varying Contract Scopes

Scope	Cybersecurity Risks	Contractual Controls
Design and build of a new facility or extension of existing facility	• Security vulnerabilities not identified and removed before delivery • New operational risks created because of inadequate design or deployment considerations	Specify technical and operational requirements. This is described later in this chapter.
Development or supply of new system or software product	• Security vulnerabilities not identified and removed before delivery • New operational risks created because of inadequate design or deployment considerations	Specify technical and operational requirements. This is described later in this chapter.
Update of existing system or software product	• New security vulnerabilities or operational risks introduced due to inadequate impact assessment • Service interruption caused by problems during update, for example a lack of backup prevents restoration of system when required	Specify technical and operational requirements. This is described later in this chapter.
System support	• Failure to follow cybersecurity management procedures introduces new security vulnerability • Failure to actively maintain products results in a cybersecurity incident exploiting an unpatched vulnerability • Service interruption caused by problems during operation, for example introduction of malware because of lack of anti-malware checks	Specify operational requirements and mandate specific policies and procedures. Specify product maintenance requirements, including patching for known security vulnerabilities.
Supply of goods	• Service interruption caused by problems during operation, for example a failure to protect system credentials gives unauthorized access to WWU equipment	Specify operational requirements and mandate specific policies and procedures.

scope involves the development or supply of a new system or software product, the following requirements will need to be incorporated in the contract. For updates to an existing system or software product, the change(s) must be assessed and some or all the following requirements will need to be incorporated:

- network security;
- software design;
- availability, reliability, maintainability, and redundancy;
- data security;
- secure remote access;
- network monitoring;
- physical access control;
- electronic access control; and
- system hardening.

These requirements can be defined or can be referenced from specific standards, in particular ISA/IEC 62443 for operational technology systems and equipment and ISO/IEC27001 for information technology systems and equipment.

3.2 Project and Operational Requirements

Several cybersecurity barriers require the following operational procedures to be followed:

- monitoring for product vulnerabilities;
- updating equipment with security patches;
- maintaining anti-malware protection;
- maintaining, testing, and restoring backups;
- managing user access; and
- managing remote access.

Third-party contracts may require contractual requirements to be added to specify adherence to these procedures. This includes during projects, as cybersecurity risks will be present well before the project deliverables are handed over to the WWU.

Monitoring for product vulnerabilities and actively updating equipment with approved vendor security patches is a common gap in WWUs (and many other organizations outside the water and wastewater sector).

It is not unusual to find WWUs running very old operating systems (such as Windows 7 or Windows XP), or very old supervisory control and data acquisition (SCADA) or PLC programming software. One challenge is ensuring compatibility across the operating system and application software. For example, some SCADA or PLC programming software may not be supported on newer versions of Windows. It is critical that operating systems and application software are actively maintained, and WWUs must ensure that this is done in-house or by a competent third party, in accordance with the operational procedures described above. This can be a challenge for very old systems in which the vendor or systems integrator no longer exists, but it creates the greatest exposure for the WWU and so must be addressed, even if this involves replacing systems, devices, or programs.

3.3 Verification of Requirements

3.3.1 Verification Processes

It is not enough to specify requirements in contracts and other documentation. There must be a verification process to confirm that the requirements have been delivered. This verification process should be captured in the contractual requirements, and will include the following:

- Documentation review—this will include verifying that the functionality in the functional specification has been correctly defined and validated, as well as verifying that all operating information, licenses, and manuals, have been provided.
- Functionality and performance testing—this will include witnessing testing to validate that the functional and performance requirements have been met.
- Visual inspection or walkthrough—this will include verifying that equipment is correctly installed and secured, and no sensitive information (such as passwords) is left on display.
- Penetration testing or Red Team assessment[1]—this may involve checking for vulnerabilities using the active network scanning or simulated exercise methods described in Chapter 3, "Assessing and Managing Cybersecurity Risk."

[1]A Red Team is a group that pretends to be an adversary. They attempt a physical or electronic intrusion in an organization at the direction of that organization, with limits on what they may or may not do. For more information, see https://csrc.nist.gov/glossary/term/red_team

3.3.2 Independent Certification

The International Society of Automation (ISA) Security Compliance Institute is a nonprofit organization that has developed several product certification programs for industrial automation and control systems and the components of these systems.[2] These programs are based on certification around the ISA/IEC 62443 Series, *Security for Industrial Automation and Control Systems*.

The current ISA Secure certification programs are listed below:

- Security Development Lifecycle Assurance (SDLA), which certifies that the security development life cycle of a vendor meets the requirements in ISA/IEC 62443-4-1, *Product Security Development Life-Cycle Requirements*.
- System Security Assurance (SSA), which certifies that products meet the requirements in ISA/IEC 62443-3-3, *System Security Requirements and Security Levels*, and have been developed in accordance with an SDLA program compliant with ISA/IEC 62443-4-1, *Product Security Development Life-Cycle Requirements*.
- Component Security Assurance, which certifies that component products meet the requirements in ISA/IEC 62443-4-2, *Technical Security Requirements for IACS Components*, and have been developed in accordance with an SDLA program compliant with ISA/IEC 62443-4-1, *Product Security Development Life-Cycle Requirements*. Certified component products can be embedded devices, such as controllers; host devices, such as PC workstations; network devices, such as firewalls; and software applications.
- IIoT Component Security Assurance, which certifies that industrial Internet of all things (IIoT) devices and IIoT gateways meet the requirements in ISA/IEC 62443-4-2, *Technical Security Requirements for IACS Components,* for at least one embedded device, host device, or network device, and have been developed in accordance with an SDLA program compliant with ISA/IEC 62443-4-1, *Product Security Development Life-Cycle Requirements*.

An additional certification program, IACS Security Assurance (IACSSA) is under development, and this will certify facilities that operate an IACS, such as a WWU operating a SCADA system.

[2]Further details of the organization and the program are at International Society of Automation. (n.d.). *ISASecure*. Retrieved January 21, 2024, from https://isasecure.org/

Certification is conducted by an independent test lab. A certificate is issued that shows details of the product, including relevant release numbers, the version of the standard referenced, and the date of certification. An example certificate is shown in Figure 6.1.

The manufacturer may use the marks:

Certified Device ISASecure

exida • CERTIFIED • IEC 62443 • SL-C 1

Certification Report:
ABB 1811109 SPC EDSA 3.0.0 ISASecure Device Assessment Report V1R0

Application Restrictions:
The unit shall be operated in a network and operational environment meeting the assumptions in the product certification report.

Validity:
Product certificate remains valid under conditions:

- The following SDLA certificate remains valid: ISASecure® SDLA certificate number ABB 1811109 C002
- issued to ABB
- The SPC Controller Version E_x remains under security management practices thereby certified

Revision 1.1 January 7, 2021

ANAB
ANSI National Accreditation Board
ACCREDITED
PRODUCT CERTIFICATION BODY

ISASecure Chartered Laboratory:
exida
80 North Main St.
Sellersville, PA 18960
License: ISCI-CL0001
ACLASS Cert No: AT-1531

exida

T-158 V2R0

Certificate / Certificat
Zertifikat / 合格証

ABB 1811109 C002

exida hereby confirms that the

SPC 600/700/800 Controller

ABB
CLEVELAND, OH USA

Has been assessed per the relevant requirements of:

ANSI/ISA-62443-4-1:2018, IEC 62443-4-1:2018 Secure product development lifecycle requirements

ANSI/ISA-62443-4-2:2018, IEC 62443-4-2:2018 Technical security requirements for IACS components

ISASecure Embedded Device Security Assurance 3.0.0 Level 1 referencing errata EDSA-102 v5.0

Meeting requirements for: Capability Security Level 1

The normative documents and issue dates that define this certification are listed at www.isasecure.org

Assessment	Subject under Assessment	Date	Current releases at time of assessment
ISASecure® EDSA evaluation	*SPC 600/700/800*	December 15, 2020	E_1

Evaluating Assessor

Certifying Assessor

FIGURE 6.1 Example ISASecure Certificate (ISASecure)

There are substantial benefits to building facilities around certified vendors and products, just as there are with hazardous-area certified equipment. These benefits are as follows:

- Products that are secure by design, developed by vendors who are certified to follow standards for their processes and procedures, form a much better foundation for a secure WWU.
- The onus is on the vendors, rather than the WWU, to obtain and maintain third-party certification for their products. The WWU can save time and money by avoiding the need to perform audits of vendors and their products. ISASecure defines and maintains the certification program that is used, and this helps ensure that third parties remain consistent and impartial in their assessments.
- To be compliant, the vendor must provide clear instruction to the WWU on secure deployment of the product. This will save additional time and money for the WWU, who will not need to develop security requirements. It also provides a greater degree of consistency that is easier to maintain.

4.0 AWARENESS

Chapter 5, "Awareness and Culture," discusses the importance of cybersecurity awareness in WWUs. This awareness is equally important for third parties. Although contracts should require that third parties provide awareness training to their employees, WWUs should additionally require that third-party employees

- take a cybersecurity induction that covers the key cybersecurity rules,
- follow a formal change-control procedure that includes updating and approving all documentation,
- take backups of equipment before and after any change,
- use protected removable media that are secured when not in use and are reformatted and virus-checked before each use, and
- scan incoming media for viruses and transfer the contents to protected removable media before being used.

As outlined in Chapter 5, "Awareness and Culture," the WWU should also track *leading* and *lagging* indicators of cybersecurity performance. In particular, the *security near misses*, which tracks the reporting of situations that could have resulted in a cybersecurity incident, should include any incidents involving third-party employees.

5.0 OVERSIGHT

5.1 Separation of Duties

Separation of duties involves ensuring that more than one person is required to complete a particular task in which safety or security might be at risk. This approach reduces the risk of fraud, theft, and human error, and is particularly useful when overseeing third parties.

Also known as the *four-eyes principle* because each process involves a "second set of eyes," this principle ensures that no single person has unilateral control or access to important processes or data, reducing the risk of fraud, mistakes, or misuse.

Typical separation of duties may involve the following:

- separate electronic authorization for critical actions, such as approval to pay invoices, or to make changes to a SCADA system configuration (e.g., alarm levels or setpoints); and
- the use of multiple security keys (physical or electronic) held by separate personnel to ensure that more than one person is aware of access to a building or system.

Cases in which it is not possible or practical to separate duties, for example in very small WWUs, alternate controls should be in place. These include

- audit trails to track who took what action, and when, and
- periodic reviews of audit trails and other records to verify all tasks are being performed as expected.

The information recorded in audit trails, and the frequency of reviews, should match the level of risk involved. It may be too late to take corrective action if the review frequency is too low. In some cases, the individuals may have left the organization, or the consequences are felt before the cause is known.

5.2 Joiners, Movers, and Leavers

When someone joins a WWU, their data and system access needs should be clear. They should have access to data and systems that are needed for them to fulfil their role, and nothing more. For example, an operator should not need access to the billing and payment system, nor should a finance administrator need access to the SCADA system. This applies equally to third parties.

There should be a formal process to arrange this access. That process should incorporate the four-eyes principle to avoid misuse.

Once someone is in a role, a periodic review process will ensure access is still required. Changes should be made with immediate effect, and records should be kept of the review and any actions arising. Accurate records allow for periodic audits to ensure the processes are being followed.

The greatest risk to any organization is posed by leavers, especially if they are disgruntled. Several of the examples in Chapter 2, "The Cybersecurity Risk to Water and Wastewater Utilities," occurred because of disgruntled former employees or contractors exploiting

- poor inventory management, such as being allowed to retain equipment after leaving;
- poor electronic account management, such as a failure to change account passwords, or remove user accounts; and
- poor anomaly and event detection, such as a failure to view access logs that show unauthorized access.

6.0 SUMMARY

Because WWUs rely heavily on third parties to enable their operations, it is imperative that strict contractual requirements are agreed. This chapter has provided a list of minimum contractual requirements that should be applied to all third-party contracts. In addition, depending on the scope of the work, it may be necessary to identify other requirements, as follows:

- Suppliers provide chemicals and other goods consumed by the WWU.
- It is imperative that cybersecurity risk management takes these third parties into account. This approach to strict management of third-party cybersecurity obligations ensures that
  - Contracts are in place, including clauses that set out the required obligations from the third party with respect to cybersecurity risk management;
  - Requirements for secure design and ongoing secure maintenance of systems and software are spelled out and executed;
  - Awareness training includes third parties, so that they are aware of the specific cybersecurity risks identified in the WWU and how they can contribute to managing that risk; and
  - Oversight of third parties is in place to ensure compliance with contractual obligations, policies, and procedures.

Chapter 7

Cybersecurity Maintenance

1.0 INTRODUCTION

Cybersecurity risk management is a continuous process because cybersecurity is constantly changing, as follows:

- cyber threats, including malware, phishing attacks, and other forms of cyberattacks, continually evolve;
- new software and hardware vulnerabilities are discovered regularly, and cybercriminals may exploit them;
- existing networks, systems, and applications require ongoing maintenance, including software patches, anti-malware updates, and backups to ensure they can be restored, as needed;
- changes to networks, systems, and applications, such as adding new hardware or software, may introduce security vulnerabilities;
- employees and contractors may join or leave, which requires changes to access requirements;
- new contracts are executed, and existing contracts updated, and the cybersecurity risks from the third parties involved must be assessed and managed; and

- laws and regulations related to cybersecurity management, data security, and privacy may change over time.

Cybersecurity risk management is not a one-time task, but a fundamental aspect of a water and wastewater utility's (WWU's) commitment to ensuring safe and secure operation and protecting customers, employees, contractors, and the environment.

Many organizations, including some WWUs, make significant investments to address cybersecurity risks, but fail to put in place the elements that are critical to managing these risks on an ongoing basis.

To manage cybersecurity risk, WWUs must institute the following ongoing activities to protect their systems, data, and operations in a constantly evolving environment:

- Incident response—Preparing for emergencies and crises, such as natural disasters or pandemics, requires ongoing planning and testing to ensure the safety of people and the environment, and the continuity of operations. Cybersecurity management includes planning for and responding to security incidents.
- Awareness and culture—Awareness training must be an ongoing activity to keep employees and third parties informed about the latest threats, best security practices, and changes to policies and procedures. Continuous tracking, reporting, and acting on cybersecurity leading indicators helps instill a culture of good cybersecurity behaviors.
- Barrier maintenance—The cybersecurity prevention and mitigation barriers may become ineffective over time if they are not actively maintained.
- Assessment and auditing—Regular assessments and audits are essential to verify that policies and procedures are being followed.
- Continuous improvement—As with any routine activity, there should be a focus on identifying ways to do things better and more efficiently. This involves continuously reviewing performance, evaluating effectiveness, and making further improvements.

Underpinning all these activities is a sound change management process. A change management process minimizes disruption, ensures consistency, mitigates risks, and maintains safety and security controls.[1]

[1]For more information about change management practices, refer to Mustard, S. (2018). *Certified mission critical operations primer* (pp. 100–104). International Society of Automation.

2.0 INCIDENT RESPONSE

2.1 Overview

Water and wastewater utilities are required to have up-to-date incident response plans to allow a response to a variety of incidents that could occur, including severe weather and loss of primary power.

Cybersecurity incidents are now commonplace, and it is not a question of if, but when a WWU will experience such an incident. As a result, it is essential that all WWUs have cybersecurity incident response plans in place.

Chapter 3, "Assessing and Managing Cybersecurity Risk," describes the approach WWUs should take to develop their incident response plan. Developing the document is only the first step in having a viable incident response plan in place. A WWU cybersecurity incident response plan must do the following:

- Consider all plausible scenarios, no matter how unlikely they may seem. For example, even though the supervisory control and data acquisition system server is not connected to the Internet, it could still be compromised by malware through other means, such as infected removable media. *Occam's razor*, a classic problem-solving statement, says that the simplest explanation is usually the best one.[2]
- Have a valid response in place for each scenario. For example, due to long lead times it may be necessary to have spare equipment available in the event of a failure or compromise of a particular device.
- Include all organizations and individuals that will need to be involved in the response, with their contact information. For example, if the programmable logic controller program is written by a system integrator out of state, it may be necessary to have them available, when needed. This will mean establishing an arrangement, in advance, with the system integrator.
- Be tested on a regular basis. This testing must be as realistic as possible, and not limited to desktop reviews. The testing should require that identified individuals take actions such as calling vendors, retrieving backups, and, for some incident scenarios, operating systems manually. This approach will ensure that any issues (such as a lack of contact details or an error in a manual procedure) can be identified and rectified before the procedure is needed to respond to a real incident.

[2]Brittanica. (n.d.). *Occam's razor*. Retrieved February 11, 2024, from https://www.britannica.com/topic/Occams-razor

To date, many organizations, including WWUs, have paid ransoms to recover their systems or data. This is due to a lack of good cybersecurity incident response preparedness. A well-prepared WWU should never need to pay a ransom to recover their system or data. In fact, paying a ransom "will not ensure that data is decrypted or that systems or data will no longer be compromised. Cybersecurity and Infrastructure Security Agency (CISA), Multi-State Information Sharing and Analysis Center, and other federal law enforcement organizations do not recommend paying ransom. In addition, attackers have begun following their ransom demands to decrypt the data with a follow-on extortion demand to keep data private."[3]

2.2 Insurance

Cybersecurity insurance policies help an organization pay for any financial losses they may incur in the event of a cyberattack or data breach. They also help them cover any costs related to the remediation process, such as paying for the investigation, legal services, and refunds to customers. Insurance companies have paid out because of several high-profile cybersecurity incidents. For example, two high-profile ransomware incidents in early 2021 were resolved when insurers negotiated ransom payments. The system owners had sufficient insurance coverage to enable payment of large sums ($4.4 million and $11 million).[4]

Whereas cybersecurity insurance may be a necessary element of a WWU's cybersecurity response, it cannot be the only element. The cybersecurity insurance market has evolved because of the growth in attacks and claims. Reports show that "[c]ompanies need to reassess their cyber insurance policies as significant breaches and growing payouts have led insurers to demand higher premiums while granting less coverage, leaving many organizations unprepared in the event of a breach or security incident."[5] As noted earlier, there is no guarantee that paying a ransom will recover a system or data anyway.

[3]Cybersecurity & Infrastructure Security Agency. (2022, March 8). *Ransomware response.* Retrieved January 21, 2024, from https://www.cisa.gov/sites/default/files/publications/Ransomware_Response_Checklist_508.pdf

[4]Mustard, S. (2022). *Industrial cybersecurity: Case studies and best practices* (pp. 230–231). International Society of Automation.

[5]Lemos, R. (2023, August 31). Insurance costs rise, coverage shrinks, but policies remain essential. *Dark Reading*. Retrieved January 21, 2024, from https://www.darkreading.com/risk/insurance-costs-rise-coverage-shrinks-but-policies-remain-essential

3.0 AWARENESS AND CULTURE

This topic is discussed in full detail in Chapter 5, "Awareness and Culture." When considering cybersecurity maintenance, WWUs must ensure that they

- provide the necessary time and resources for employees to perform their roles;
- provide the right tools to allow them to do this securely and safely, while still being efficient; and
- create a culture in which cybersecurity is a routine part of everyone's job, not an optional extra task to do if time permits.

Regular awareness training is essential to ensure that employees and third parties are informed about the latest threats, best security practices, and changes to policies and procedures.

Additional routine tests should be performed, such as phishing tests and clear workspace checks. This will not only remind employees and third parties of their responsibilities, but it will also highlight that cybersecurity management is a priority for the WWU.

Unannounced spot-checks should also be performed periodically to ensure that everyone remains diligent. These checks might include trying to use unsecure removable media or requesting another user's credentials to gain access to a system.

Water and wastewater utilities should also track cybersecurity leading indicators such as training completions, phishing test results, clear workspace checks, and near-miss incidents; report the results to employees and third parties; and take action to correct any negative trends in cybersecurity behaviors before a serious cybersecurity incident occurs.

4.0 BARRIER MAINTENANCE

4.1 Overview

As noted in Chapter 3, "Assessing and Managing Cybersecurity Risk," the protective barriers, or controls, that are in place to manage cybersecurity can develop holes or vulnerabilities over time, for example:

- new software and hardware flaws are discovered;
- changes are made to networks, systems, or applications;
- employee and contractors join, leave, and change role; and
- policies and procedures are not followed.

The activities shown in Table 7.1 must be part of routine operations to minimize the number of barrier holes or vulnerabilities.

4.2 Procedures

The list of activities in Section 4.1 may seem onerous, especially to a WWU with limited resources. One way to address this concern is to document the procedures that are needed to undertake these activities. Documenting the procedures will help by

- clarifying what is required,
- allowing allocation of specific tasks to specific resources,
- providing the basis for training of employees and third parties,

TABLE 7.1 Maintenance Activities for Cybersecurity Barriers

Barrier	Maintenance Activities
Prevention	
Secure design	• Ensure that all changes are reviewed and approved by a competent person, and any new risks are managed. • Ensure that all design documentation is kept up to date.
Secure remote access	• Ensure that remote access is limited to approved users, systems, and methods. • Maintain lists of approved users and update, as needed, for new joiners, leavers, or changes in role.
Network monitoring	• Ensure that monitoring is operating and up to date. • Ensure that alerts and logs are being reviewed on a regular basis to identify abnormal behavior as early as possible.
Physical access control	• Ensure that physical access controls (locks, gates, doors, etc.) are in place. • Maintain lists of approved users with access/keys and update, as needed, for new joiners, leavers, or changes in role.
Electronic access control	• Maintain lists of approved users with their system/ application access and update, as needed, for new joiners, leavers, or changes in role.
System hardening	• Ensure that operating system, application software, and device firmware is maintained with the latest vendor-approved security patches. • Ensure that anti-malware tools are maintained and virus signatures are kept up to date. • Ensure that system hardening controls (universal serial bus [USB] locks, Wi-Fi security, etc.) are in place.

(*continued*)

TABLE 7.1 Maintenance Activities for Cybersecurity Barriers (*Continued*)

Barrier	Maintenance Activities
Supervision and communications	• Ensure that product vulnerabilities are tracked. • Ensure that training is provided to new joiners (employees or third parties). • Ensure that training is updated to reflect any changes in threats, vulnerabilities, or processes and procedures. • Monitor cybersecurity behavior leading indicators and act, as needed. • Audit compliance against policies and procedures.
Mitigation	
Cybersecurity incident response plan	• Ensure that the response plan is tested regularly, and any updates are made following the exercise. • Ensure that the plan is updated to reflect any changes in contacts or procedures.
Backup and recovery procedures	• Ensure that backups are being maintained as defined in procedures. • Ensure that restore processes are tested regularly.
Mechanical fail-safes	• Ensure that mechanical fail-safes (alarms, valves, independent sensors, etc.) are operational and will function when needed.
Manual procedures	• Ensure that relevant employees and third parties are trained in procedures for manual operation (e.g., operating a water treatment facility without automatic control). • Test knowledge of relevant employees and third parties with regular exercises.

- providing the basis for auditing activities,
- improving consistency and repeatability,
- minimizing the opportunity for error, and
- providing the basis for identifying improvements.

The documentation need not be onerous. As a minimum, procedures should include

- a unique identification, such as title and reference number;
- a version number and history of revisions;
- any prerequisites before undertaking the procedure;
- any specific competencies, tools, or equipment that is required; and
- a step-by-step description of the procedure.

Procedure descriptions can be defined in table or flowchart format. An example of a simple procedure, to perform a backup, is shown in Table 7.2 and in flowchart form in Figure 7.1. These examples show how different roles can be identified in the procedure.[6]

Table 7.3 presents a list of recommended procedures that WWUs should document.

4.3 Documentation

Accurate documentation is essential for safe, secure, and reliable operations. The following is a list of typical documentation that should be available for key systems used by WWUs. Table 7.4 defines the typical updates that these documents will need to continue to be effective.

The change control process described in Section 4.2 should include a check that all relevant documentation is updated because of a change.

There should also be periodic checks that the change control process is being followed and documentation is being updated.

5.0 ASSESSMENT AND AUDITING

Water and wastewater utilities should conduct regular assessments and audits to verify that

- all cybersecurity risks are quantified correctly,
- all cybersecurity scenarios are covered in incident response plans,

TABLE 7.2 Backup Procedure in Tabular Form

Step	Description	Role A	Role B	Role C
1	Remove backup media from storage facility	X		
2	Insert backup media into drive X		X	
3	Execute backup program		X	
4a	On successful completion, remove backup media from drive X		X	
4b	If unsuccessful, report to Role A to resolve reported issue		X	
5	Return backup media to storage facility	X		
6	Update backup log			X

[6]Mustard, S. (2018). *Certified mission critical operations primer* (pp. 83–86). International Society of Automation.

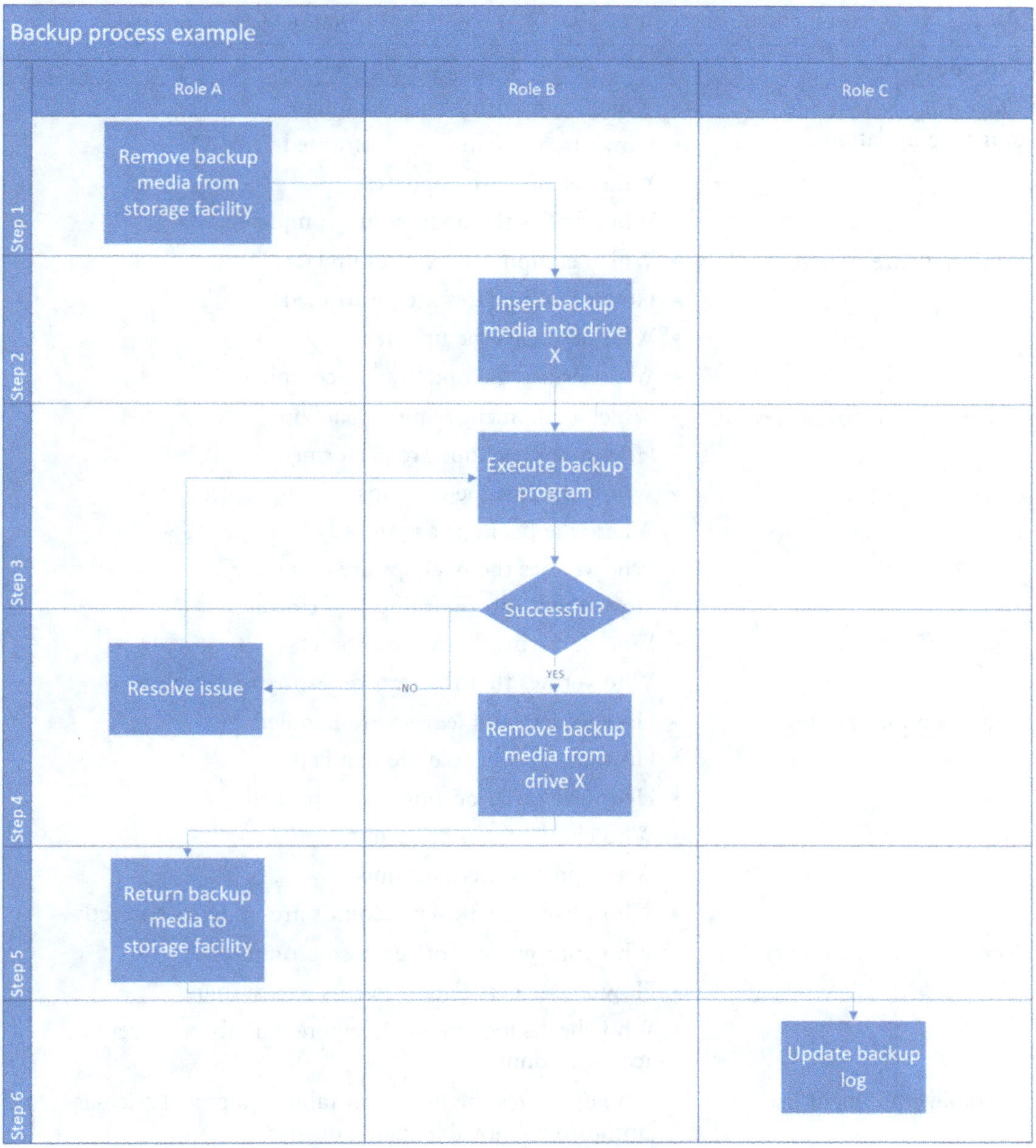

FIGURE 7.1 Backup Procedure in Flowchart Form

TABLE 7.3 Recommended Procedures That WWUs Should Document

Procedure	Includes
Operating system and software updates	• Which equipment requires update • How often updates are performed • Who performs the updates • Who verifies the updates are complete
Anti-malware updates	• Which equipment requires update • How often updates are performed • Who performs the updates • Who verifies the updates are complete
Backup and recovery testing	• Which equipment requires backup • How often backups are performed • Who performs the backups • Where the backups are stored • Who verifies the backups are complete • How often restore testing is performed • Who performs the restore testing • Who verifies that the restore testing is complete
User account updates	• How joiners and leavers are handled • How changes in role are handled • How often user accounts are checked • Who checks the user accounts • Who updates user accounts • Who verifies that user accounts are updated correctly
Remote access management	• Who approves remote access accounts • How a user gets remote access to a system • Who checks logs to identify potential abuse of remote access account
Portable equipment use	• Who approves the use of portable equipment such as laptops or removable media on-site • What checks must be performed before using portable equipment on-site
Change control and review	• Who reviews proposed changes to networks, systems, or applications • Who checks that documentation is updated to reflect the change
Monitoring system and log file analysis	• Who reviews monitoring systems and log files • How often are monitoring systems and log files reviewed • What action is performed based on results of review

(*continued*)

TABLE 7.3 Recommended Procedures That WWUs Should Document (*Continued*)

Procedure	Includes
Manual operation	• What manual operations may be required in the event of a loss of automated control systems • Who performs the manual operations • Who is contacted • How often are exercises conducted • Who verifies that updates to the manual operation procedures are made because of exercises
Incident and near-miss reporting	• How are incidents and near misses reported • Who maintains the data • What actions are performed because of incidents or near misses

TABLE 7.4 Required Document Updates During Ongoing Maintenance

Document	Purpose
Functional specification	This document should be updated if any changes are made to the deliverable(s).
Test specification	This document should be updated if any changes are made to the deliverable(s).
Network diagram	This diagram should be updated if any changes are made to the deliverable(s).
Physical layout drawings	These drawings should be updated if any changes are made to the deliverable(s).
Wiring diagrams and schedules	These documents should be updated if any changes are made to the deliverable(s).
Inventory list	This list should be updated if any changes are made to the deliverable(s).
Operations manual	This document should be updated if any changes are made to the deliverable(s).

- all policies and procedures are correct and are being followed, and
- data shows that cybersecurity is being managed effectively.

Assessments include

- risk assessment,
- gap assessment, and
- incident response test.

An audit is a systematic evaluation of a WWU's policies, procedures, and processes to determine their effectiveness in safeguarding against a cybersecurity incident.

Audits are typically conducted by independent groups, either from within the WWU (for instance, a city or local government resource) or externally (for instance, commercial organizations specializing in cybersecurity management or from a separate government body such as CISA or the U.S. Environmental Protection Agency). This ensures an unbiased view of the WWU's cybersecurity preparedness.

6.0 CONTINUOUS IMPROVEMENT

Water and wastewater utilities should always look to improve their cybersecurity preparedness. Results from assessments, audits, near misses, incidents, and incident response exercises should be used, in conjunction with current standards and guidance, to review and update processes and procedures to

- close any gaps in networks, systems, or applications;
- close any gaps in processes or procedures;
- better respond to future incidents;
- reduce the number of near misses; and
- streamline the processes and procedures without reducing their effectiveness.

7.0 SUMMARY

Cybersecurity management is a continuous process. It involves adapting to new cybersecurity threats and vulnerabilities, new technology, and changes in regulatory requirements. Water and wastewater utilities need to establish continuous activities relating to the following:

- Incident response—Planning for and responding to security incidents.
- Awareness and culture—Ensuring employees and third parties are informed about the latest threats, best security practices, and changes to policies and procedures, as well as continuous tracking, reporting, and acting on cybersecurity leading indicators.
- Barrier maintenance—Ensuring that the cybersecurity prevention and mitigation barriers remain effective over time.

- Assessment and auditing—Verifying that policies and procedures are being followed.
- Continuous improvement—Continuously reviewing performance, evaluating effectiveness, and making further improvements.

Chapter 8

Terminology, Resources, and Further Reading

1.0 INTRODUCTION

There is a great deal of guidance on cybersecurity, so much that it can be hard to know where to start. This chapter combines a list of common terminology, key resources, and further reading to help provide some guidance.

2.0 TERMINOLOGY

Table 8.1 contains terms and abbreviations that are either in this book or may be found in some of the resources and further reading referenced here.

TABLE 8.1 Terms and Abbreviations

Term or Abbreviation	Definition
AAA	Authentication, Authorization, and Accounting
ACL	Access Control List
AAD	Azure Active Directory
AD	Active Directory
ALARP	As Low As Reasonably Practicable, a definition of the point where risk is being managed as well as possible
APT	Advanced Persistent Threat
ANSI	American National Standards Institute
AV	Antivirus
Availability	The probability that the system is operating properly when it is required. Availability is measured as a percentage over a defined period
AWIA	America's Water Infrastructure Act
AWWA	American Water Works Association
Barrier	A technical or procedural element that minimizes the likelihood or consequence of an incident. Also known as a control
Bowtie	A diagrammatic method used to show the relationship between threats, causes, consequences, controls, and escalation factors
BYOD	Bring Your Own Device
CCTV	Closed-Circuit Television
CERT	Computer Emergency Response Team
CIA	Confidentiality, Integrity, Availability (security triad)
CISA	Cybersecurity and Infrastructure Security Agency
CMMS	Computerized Maintenance Management System
Consequence	The result of an incident. For instance, ransomware on a computer system can result in a loss of business operations
Control	Another term for barrier, a technical or procedural element that minimizes the likelihood or consequence of an incident
CPG	Cybersecurity Performance Goal
CSIRT	Computer Security Incident Response Team
CVE	Common Vulnerabilities and Exposures
DDoS	Distributed Denial of Service
DHCP	Dynamic Host Configuration Protocol

(continued)

TABLE 8.1 Terms and Abbreviations (*Continued*)

Term or Abbreviation	Definition
DHS	Department of Homeland Security
DLP	Data Loss Prevention
DMZ	Demilitarized Zone
DNS	Domain Name System
EDR	Endpoint Detection and Response
Encryption	A method of protecting sensitive information by converting it into an unreadable form
EPP	Endpoint Protection Platform
Escalation Factor	A way in which the effectiveness of a barrier or control can be degraded
Exploit	A means to use a vulnerability to create a cybersecurity incident
Fail-safe	A type of barrier or control that is in place to minimize the consequences of an incident. For example, a pressure relief valve
FBI	Federal Bureau of Investigation
FIDO	Fast Identity Online, an open standard and a set of technology specifications designed to improve online authentication and enhance digital security
Firewall	Network Security Device
Four-eyes	A practice that requires the involvement of at least two individuals in a critical action or decision. It is a control mechanism to reduce the risk of fraud or errors
GIS	Geospatial Information System
Geofencing	Geofencing uses location data to define boundaries that can be used to trigger actions, notifications, or apply policies
GPS	Global Positioning System
Hash	A unique representation of data that can be used to validate that data is not compromised during transmission
HIDS/NIDS	Host-based Intrusion Detection System/Network-based Intrusion Detection System
HiPo	High Potential, a type of incident (usually safety but can apply to cybersecurity)
HTTP	Hypertext Transfer Protocol
HTTPS	Hypertext Transfer Protocol Secure

(*continued*)

TABLE 8.1 Terms and Abbreviations (*Continued*)

Term or Abbreviation	Definition
IACS	Industrial Automation and Control System
IAM	Identity and Access Management
ICS	Industrial Control System
IDS	Intrusion Detection System
IEC	International Electro Technical Commission
IoT	Internet of Things
IPS	Intrusion Prevention System
ISA	International Society of Automation
ISO	International Standards Organization
IT	Information Technology
Key	Data used in an encryption algorithm, combined with the original data to make the encrypted form
KPI	Key Performance Indicator
LDAP	Lightweight Directory Access Protocol
LAN	Local Area Network
Likelihood	The probability that an event will occur
Maintainability	The ease with which a product can be maintained and is an essential element for successful operations
MAC	Media Access Control
MDM	Mobile Device Management
MFA	Multifactor Authentication
MITM	Man-in-the-Middle
MS-ISAC	Multi-State Information Sharing and Analysis Center, a U.S.-based organization that focuses on enhancing the cybersecurity resilience of state, local, tribal, and territorial (SLTT) governments
NAC	Network Access Control
NIST	National Institute of Standards and Technology
NRWA	National Rural Water Association
NTP	Network Time Protocol
OAUTH	Open Authorization, an open standard authentication and authorization framework
OSINT	Open-Source Intelligence
OT	Operational Technology
PCI DSS	Payment Card Industry Data Security Standard
PII	Personally Identifiable Information

(*continued*)

TABLE 8.1 Terms and Abbreviations (*Continued*)

Term or Abbreviation	Definition
PIN	Personal Identification Number
PKI	Public Key Infrastructure
PLC	Programmable Logic Controller
PoLP	Principle of Least Privilege
RAT	Remote Access Trojan
RBAC	Role-Based Access Control
RCE	Remote Code Execution
Redundancy	Achieving high availability by replicating hardware so that if one device fails, another can take over
Red Team	A red team is a group that pretends to be an adversary. They attempt a physical or electronic intrusion in an organization at the direction of that organization, with limits on what they may or may not do. For more information, see https://csrc.nist.gov/glossary/term/red_team
Reliability	The probability of a component or system to perform a required function in a specific environment for a specific period without failure
Risk	The result of exposure to a hazard, or threat. Risk is typically determined from the product of likelihood and consequence.
RTU	Remote Telemetry Unit or Remote Terminal Unit
SCADA	Supervisory Control and Data Acquisition
SIEM	Security Information and Event Management
SOP	Standard Operating Procedure
SSO	Single Sign-On
Threat	Also known as a hazard, it could be environmental-based, such as flooding or earthquake, or infrastructure-based, such as loss of power or communications, or others, such as cybersecurity-related (hacker, nation-state, etc.)
TLS/SSL	Transport Layer Security/Secure Sockets Layer
U.S. EPA	U.S. Environmental Protection Agency
USB	Universal Serial Bus
VLAN	Virtual Local Area Network
VPN	Virtual Private Network

(*continued*)

TABLE 8.1 Terms and Abbreviations (*Continued*)

Term or Abbreviation	Definition
Vulnerability	A flaw in software or hardware, a configuration error, or a failure in a process or procedure
Water-ISAC	Water Information Sharing and Analysis Center, an organization that focuses on enhancing the cybersecurity and physical security resilience of the water and wastewater sector in the United States
WEF	Water Environment Federation
WEP	Wired Equivalent Privacy
WPA	Wi-Fi Protected Access
WWV	NIST radio station that broadcasts time and frequency information 24 hours per day, 7 days per week
Zero Day	Vulnerability exploited before it is known to the vendor

3.0 RESOURCES

3.1 *America's Water Infrastructure Act* Risk and Resilience Assessments and Emergency Response Plans

America's Water Infrastructure Act (AWIA) Section 2013 requires community (drinking) water systems (CWSs) serving more than 3,300 people to develop or update risk and resilience assessments (RRAs) and emergency response plans (ERPs). The law specifies the components that the RRAs and ERPs must address and establishes deadlines by which water systems must certify to U.S. Environmental and Protection Agency (U.S. EPA) completion of the RRA and ERP.

Each community water system serving a population of greater than 3,300 persons shall assess the risks to, and resilience of, its system. Such an assessment shall include

- the risk to the system from malevolent acts and natural hazards;
- the resilience of the pipes and constructed conveyances, physical barriers, source water, water collection and intake, pretreatment, treatment, storage and distribution facilities, electronic, computer, or other automated systems (including the security of such systems) that are used by the system;
- the monitoring practices of the system;
- the financial infrastructure of the system;

- the use, storage, or handling of various chemicals by the system; and
- the operation and maintenance of the system.

The assessment may include an evaluation of capital and operational needs for risk and resilience management for the system.[1]

The assessment methodology described in this book is aimed at cybersecurity risks ("electronic, computer, or other automated systems [including the security of such systems], which are utilized by the system"), but can be extended to address the other risks identified.

3.2 National Institute of Standards and Technology Cybersecurity Framework

U.S. Presidential Executive Order 13636 ("Improving Critical Infrastructure Cybersecurity") instructed the National Institute of Standards and Technology (NIST) to develop a voluntary cybersecurity framework (CSF) that would provide a "prioritized, flexible, repeatable, performance-based, and cost-effective approach for assisting organizations responsible for critical infrastructure services to manage cybersecurity risk."[2]

Version 2.0 of the CSF is structured into six core functions, each of which includes categories and subcategories. This format enables those unfamiliar with the requirements of cybersecurity management to navigate the subject and drill into detail, as needed.[3]

These six core functions identify the following high-level activities that organizations such as water and wastewater utilities (WWUs) should do:

1. Govern—Define roles and responsibilities, assess. Risk, create and maintain policies, and perform oversight.
2. Identify—Create an inventory of all cyber assets and understand what these assets do, what hardware and software is used, how the equipment is connected, what external connections exist, and so on.
3. Protect—Create the defense in-depth protections, such as access control policies and procedures, malware protection and patching procedures, and backup and recovery strategies.

---

[1]U.S. Environmental and Protection Agency. (n.d.). *RRA requirements and assistance resources for CWSs that serve more than 3,300*. Retrieved November 5, 2023, from https://www.epa.gov/waterresilience/awia-section-2013#RRA

[2]Obama Whitehouse Archives. (n.d.). *Cybersecurity – Executive order 13636*. Retrieved February 11, 2024, from https://obamawhitehouse.archives.gov/issues/foreign-policy/cybersecurity/eo-13636

[3]National Institute of Standards and Technologies. (n.d.). *Cybersecurity framework*. Retrieved February 11, 2024, from https://www.nist.gov/cyberframework

4. Detect—Put in place methods, automatic and manual, to detect unusual activity or unauthorized access.
5. Respond—Establish procedures to deal with cybersecurity events.
6. Recover—Ensure that procedures are in place before and after cybersecurity events to ensure that it is possible to restore normal operation.

The CSF overview is illustrated in Figure 8.1. It shows the six core functions—Govern, Identify, Protect, Detect, Respond, and Recover—with their respective categories (e.g., Asset Management, Identity Management, Authentication, and Access Control).

As noted previously, as a framework, the CSF does not provide any detailed guidance. Instead, the document refers to standards and guides. This format helps readers who are unfamiliar with the standards and guides to navigate the documents.

The methodology described in this book takes all five core functions into account, as shown in Table 8.2.

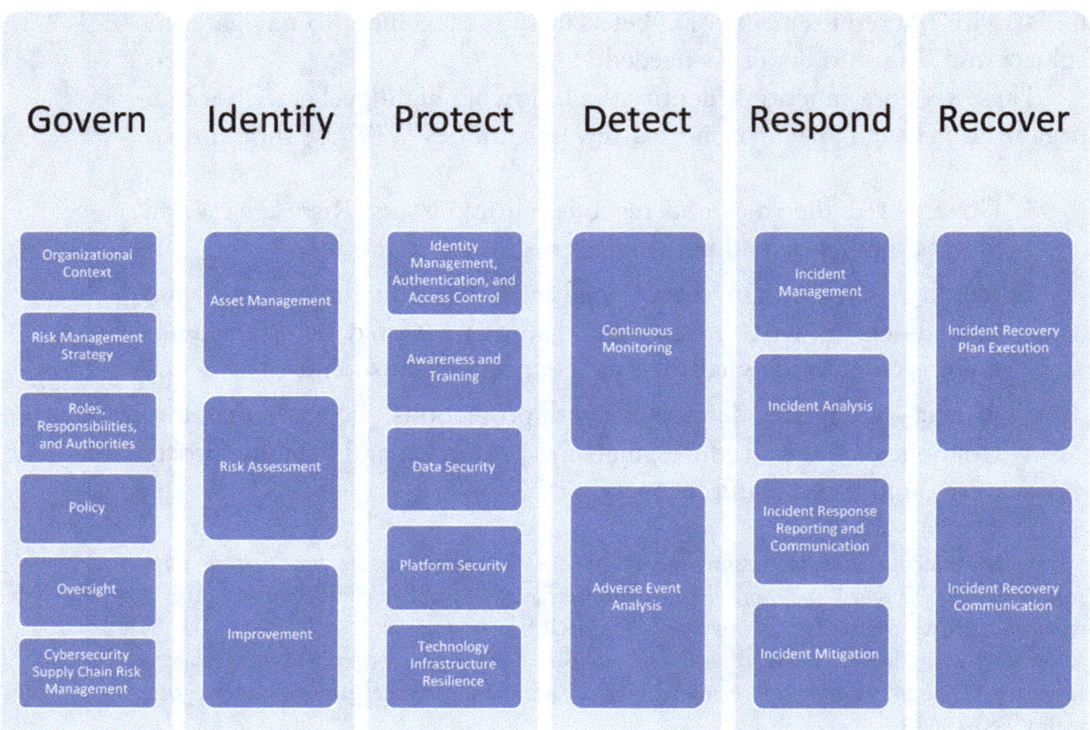

FIGURE 8.1 NIST Cybersecurity Framework

TABLE 8.2 NIST Core Functions Mapped to This Book's Chapters

National Institute of Standards and Technology CSF Core Function	Methodology Addressing Core Function
Govern	Quantifying risk (Chapter 3)
	Policy, roles and responsibilities, and oversight (Chapter 4)
	Dealing with third parties (Chapter 6)
Identify	Risk assessment (Chapter 3)
	Documentation (Chapter 4)
	Improvement (Chapter 7)
Protect	Secure remote access (Chapter 3 & 4)
	Physical access control (Chapter 3 & 4)
	Electronic access control (Chapter 3 & 4)
	System hardening (Chapter 3 & 4)
Detect	Network monitoring (Chapter 4)
Respond	Supervision and communications (Chapter 3 & 4)
	Cybersecurity incident response plan (Chapter 3 & 4)
	Backup and recovery procedures (Chapter 3 & 4)
	Mechanical failsafes (Chapter 3 & 4)
	Manual procedures (Chapter 3 & 4)
Recover	Cybersecurity incident response plan (Chapter 3 & 4)
	Supervision and communications (Chapter 3 & 4)

3.3 International Society of Automation/International Electro Technical Commission 62443

The International Society of Automation (ISA)/International Electro Technical Commission (IEC) 62443 series of standards addresses the security of industrial automation and control systems (IACSs) such as supervisory control and data acquisition (SCADA) systems used by WWUs.

Figure 8.2 shows an overview of the family of standards.

There are four tiers in the series of standards. The first two focus on people and processes. The last two focus on technology (systems and components). Key documents in the family include the items listed in Table 8.3.

The series of standards defines seven foundational requirements (FRs) that organize the technical requirements for cybersecurity management of industrial automation and control systems. These are detailed in ISA/

FIGURE 8.2 The ISA/IEC 62443 Series of Standards

IEC62443-3-3, System Security Requirements and Security Levels. As noted above, in the context of a WWU, the term *IACS* refers to a SCADA system or equivalent, but the requirements can also be applied to other operational technology systems in the WWU as follows:[4]

- **FR 1, Identification and Authentication Control (IAC)**—Identify and authenticate all users (humans, software processes, and devices) before allowing them access to the IACS.
- **FR 2, Use Control (UC)**—Enforce the assigned privileges of an authenticated user (human, software process, or device) to perform the requested action on the IACS and monitor the use of these privileges.
- **FR 3, System Integrity (SI)**—Ensure the integrity of the IACS to prevent unauthorized manipulation.

[4]International Society of Automation. (n.d.). *ISA/IEC62443-3-3-2013, Security for industrial automation and control systems part 3-3: System security requirements and security levels.*

TABLE 8.3 List of Key ISA/IEC62443 Mapped to This Book's Chapters

ISA/IEC62443	Title and Purpose	Relevant Chapter(s)
-2-1	Establishing an IACS security program. This helps organizations plan and implement a cybersecurity management system focused on industrial cybersecurity.	The requirements of the IACS security program are • Risk Assessment (Chapter 3) • Implement Controls (Chapter 4) • Maintenance (Chapter 7) • Policies, procedures, training, and awareness (Chapters 4, 5, and 6) • Audits (Chapter 7)
-3-2	Security risk assessment, system partitioning, and security levels. Defines the requirements for addressing the cybersecurity risks in an IACS, including the use of zones and conduits as well as security levels. These are key aspects of industrial cybersecurity design.	This is described in Chapters 3 and 4.
-3-3	System security requirements and security levels. Defines the requirements for an IACS system based on a specified security level. It helps organizations quantify their requirements in universally understood terms.	This is described in Chapter 4.
-4-1	Product security development life cycle requirements. Defines the requirements for a product developer's security development life cycle.	This applies to product vendors and is described in Chapter 6.
-4-2	Technical security requirements for IACS components. Defines the requirements for IACS components based on the required security level. Components include devices and applications.	This applies to product vendors and is described in Chapter 6.

- **FR 4, Data Confidentiality** (DC)—Ensure the confidentiality of information on communication channels and in data repositories to prevent unauthorized disclosure.
- **FR 5, Restricted Data Flow** (RDF)—Segment the control system via zones and conduits to limit the unnecessary flow of data.

- **FR 6, Timely Response to Events (TRE)**—Respond to security violations by notifying the proper authority, reporting needed evidence of the violation, and taking timely corrective action when incidents are discovered.
- **FR 7, Resource Availability (RA)**—Ensure the availability of the control system against the degradation or denial of essential services.

The barriers described in this book are shown in the Table 8.4 together with the relevant ISA/IEC62443 FR(s). Using this cross-reference and the results from the gap assessment described in Chapter 3, "Assessing and Managing Cybersecurity Risk," it is possible to determine the Achieved Security Level (SL-A) for the WWU, if desired.

As noted in Chapter 3, "Assessing and Managing Cybersecurity Risk," the methodology in this book focuses on helping WWUs achieve SL 1 throughout their operational environment, with some SL 2, SL 3, and SL 4 capabilities where possible (e.g., use of multifactor authentication, or MFA, in user authentication and authorization). The methodology can also be used in conjunction with this cross-reference to identify the specific system requirements needed to achieve a particular security level, if required.

TABLE 8.4 Cybersecurity Barriers and Relevant ISA/IEC62443 FR(s)

Barrier	Technical	Procedural	ISA/IEC 62443 FR Reference
Prevention Barriers			
Secure design	X	X	FR 4, FR 5
Secure remote access	X	X	FR 1, FR 2
Network monitoring	X	X	FR 6, FR 7
System hardening	X	X	FR 3, FR 7
Physical access control	X	X	FR 1
Electronic access control	X	X	FR 1, FR 2
Mitigation Barriers			
Cybersecurity incident response plan	X	X	FR 6
Backup and recovery procedures	X	X	FR 6, FR 7
Mechanical failsafes	X	X	FR 7
Manual procedures		X	FR 6

3.4 International Standards Organization/International Electro Technical Commission 27001

International Standards Organization/IEC 27001 is an international standard for information security management systems. The standard is designed to help organizations protect the confidentiality, integrity, and availability of their information assets.

3.5 Center for Internet Security Critical Security Controls

The Center for Internet Security (CIS) Critical Security Controls is a set of best practices and guidelines for cybersecurity resilience. The controls provide a prioritized and actionable framework for managing cybersecurity

TABLE 8.5 ISO/IEC 27001 Requirement Areas Mapped to This Book's Chapters

Requirement Area	Purpose	Relevant Chapter(s)
Risk Management	Requirements to identify and assess information security risks and then implement appropriate controls to mitigate those risks.	The risk management methodology is described in Chapter 3. The details of information security risks are described in Chapter 4.
Policies and Procedures	Requirements for policies, procedures, and guidelines to govern information security practices within the organization.	Described in Chapter 4.
Asset Management	Requirements to identify and classify information assets, ensuring they are adequately protected.	Described in Chapter 4.
Access Control	Requirements for controlling access to information and information processing facilities, including user access, network access, and physical access.	Described in Chapter 4.
Incident Response	Requirements for incident response planning to handle and respond to security incidents effectively.	Described in Chapter 4.
Compliance and Legal Obligations	Requirements to comply with legal and regulatory requirements related to information security and privacy.	Compliance and legal obligations are considered during the risk assessment process.
Continuous Improvement	Requirements for regular monitoring, review, and updating of information security controls and practices.	Described in Chapter 7.

risks. The methodology described in this book takes all 18 controls described in CIS CSC version 8 into account, as summarized in Table 8.6.[5]

3.6 Cross-Sector Cybersecurity Performance Goals

The Cybersecurity and Infrastructure Security Agency's (CISA's) Cybersecurity Performance Goals (CPGs) are a subset of cybersecurity practices, selected through a thorough process of industry, government, and expert consultation, aimed at meaningfully reducing risks to both critical infrastructure operations and the American people. These voluntary CPGs strive to help small- and medium-sized organizations kickstart their cybersecurity efforts by prioritizing investment in a limited number of essential actions with high-impact security outcomes.

TABLE 8.6 CIS Controls Mapped to This Book's Guidance

Number	Control	Included in this book's guidance?
1	Inventory and Control of Enterprise Assets	√
2	Inventory and Control of Software Assets	√
3	Data Protection	√
4	Secure Configuration of Enterprise Assets and Software	√
5	Account Management	√
6	Access Control Management	√
7	Continuous Vulnerability Management	√
8	Audit Log Management	√
9	Email and Web Browser Protections	√
10	Malware Defenses	√
11	Data Recovery	√
12	Network Infrastructure Management	√
13	Network Monitoring and Defense	√
14	Security Awareness and Skills Training	√
15	Service Provider Management	√
16	Application Software Security	√
17	Incident Response Management	√
18	Penetration Testing	√

[5] The full details of each control can be obtained from the CIS website (https://www.cisecurity.org/). Readers should always consult the CIS website to ensure they obtain the latest version of the controls.

The CPGs are intended to be

- a baseline set of cybersecurity practices broadly applicable across critical infrastructure with known risk-reduction value;
- a benchmark for critical infrastructure operators to measure and improve their cybersecurity maturity;
- a combination of recommended practices for information technology and operational technology owners, including a prioritized set of security practices; and
- unique from other control frameworks as they consider not only the practices that address risk to individual entities, but also the aggregate risk to the nation.

The CISA's CPGs have been organized to align to the NIST CSF functions.[6]

As a result of the alignment with the NIST CSF, WWUs can use the methodology described in this book to meet CISA's CPGs.

3.7 NIST Special Publication 800 Series

The NIST Special Publication 800 series is a series of reports on NIST's information technology laboratory research, guidelines, and outreach efforts in information system security. Included in this series is NIST Special Publication 800-82, *Guide to Operational Technology (OT) Security*. This document provides guidance on how to secure operational technology, including SCADA systems used by WWUs. The document provides an overview of ICS and typical system topologies. It identifies typical threats and vulnerabilities to these systems and provides recommended security countermeasures to mitigate the associated risks.

The ISA/IEC62443 series of standards covers all the recommendations from NIST 800-82. In addition, ISA/IEC62443 and NIST 800-82 are referenced in the NIST CSF. As a result, the methodology described in this book covers all recommendations in NIST 800-82.

3.8 American National Standards Institute/American Water Works Association Risk Management Standards

The American Water Works Association (AWWA) produces standards, manuals, and other publications giving guidance on risk management. The following AWWA risk management standards provide details on the management of cybersecurity risks alongside other risks:

[6]Taken verbatim from the CISA website. Retrieved November 5, 2023, from https://www.cisa.gov/cross-sector-cybersecurity-performance-goals

- ANSI/AWWA G430-14(R20): Security Practices for Operation and Management. Covers the minimum requirements for a protective security program for a water, wastewater, or reuse utility. Information protection and continuity are requirements.
- ANSI/AWWA G440-17: Emergency Preparedness Practices. Covers minimum requirements to establish and maintain an acceptable level of emergency preparedness based on identified and perceived risks facing water utilities. Key business considerations and operating system recovery.
- ANSI/AWWA J100-21: Risk and Resilience Management of Water and Wastewater Systems. Enables water and wastewater utility owners and operators to make sound decisions when allocating limited resources to reducing risk and improving resilience. Cyber is a required threat domain.
- ANSI/AWWA G300-14: Source Water Protection. Critical requirements for the effective protection of source waters.

The AWWA aligns its cybersecurity guidance with the NIST CSF. As such, the methodology described in this book covers all cybersecurity risk management recommendations from AWWA.

3.9 Assessment Tools

The following freely available tools are available to help WWUs conduct cybersecurity assessments.

Vulnerability Self-Assessment Tool (VSAT) for water and wastewater risk assessment (https://www.epa.gov/waterriskassessment/vulnerability-self-assessment-tool-conduct-drinking-water-or-wastewater-utility)

U.S. EPA recommends that the quantitative risk assessment be limited to a small number of the assets and threats that present the highest risk. Attempting to analyze a high number of asset/threat pairs in a single assessment will render the assessment difficult to complete and make the interpretation of results challenging. Strategies to reduce the number of asset/threat pairs including grouping similar assets into a single asset category and limiting threats to those with the highest likelihood of occurrence.

Cyber Security Evaluation Tool (CSET)

(https://www.cisa.gov/downloading-and-installing-cset)

The CSET is a desktop software tool that guides asset owners and operators through a step-by-step process to evaluate industrial control system (ICS) and information technology (IT) network security practices. Users can evaluate their own cybersecurity stance using many recognized government and industry standards and recommendations. CSET v11.5 includes a Cyber Performance Goals (CPG) Assessment.

AWWA Cybersecurity Risk Management Tool
(https://cybersecurity.awwa.org/)

The AWWA Water Sector Cybersecurity Risk Management Tool is an online tool to support utilities in developing their cybersecurity risk management strategy. The tool is aligned with the NIST CSF.

These tools incorporate questionnaires to help collect information to identify gaps in the cybersecurity barriers of the WWU. As with all questionnaires, it is important to ensure that the correct questions are asked, to get the right answers. When completing any questionnaire, the WWU should check that questions address all barriers identified in Chapter 4, Cybersecurity Barriers. The methodology described in this book is not dependent on any of these tools, but the tools can be used by WWUs as part of the methodology, if desired.

3.10 Other Resources

The following resources are highly recommended to WWUs. Cybersecurity management is a continuous activity and WWUs will benefit from connecting and collaborating with these sources of knowledge and support.

National Rural Water Association—NRWA (https://nrwa.org/issues/cybersecurity)

The National Rural Water Association is a nonprofit organization dedicated to training, supporting, and promoting the water and wastewater professionals that serve small and rural communities across the country.

The NRWA focuses on simple guidance and practical steps that will help all members better manage their cybersecurity risk:

The NRWA is developing a plan for a comprehensive continuous cyber education program that will help all small and rural water and wastewater systems better manage their cybersecurity risk, by

- Creating a multi-stage training program that targets guidance based on risk along with the technical ability and capacity of systems;
- Focusing on training Circuit Riders, the key NRWA field experts who are best placed to sustain the ongoing effort needed to manage the cybersecurity risk; and
- Integrating with other initiatives around water sector cybersecurity to ensure that time, effort, and funds are used to best effect.

Water Information Sharing and Analysis Center—WaterISAC (https://www.waterisac.org)

WaterISAC is the designated information sharing and operations arm of the Water Sector Coordinating Council.

WaterISAC has established secure and close contact with partners in government to access sensitive and classified security information. WaterISAC maintains two-way communication with the CISA, the Federal Bureau of Investigation (FBI), the U.S. EPA, state intelligence fusion centers, and other federal and state agencies.

Through the National Council of ISACs, WaterISAC also collaborates with more than two dozen other ISACs, particularly with the Electricity ISAC. They also amass intelligence and resources from selected security experts, water and wastewater industry partners, news media, and their own members.

American Water Works Association—AWWA (https://www.awwa.org/Resources-Tools/Resource-Topics/Risk-Resilience/Cybersecurity-Guidance)

The AWWA is an international, nonprofit, scientific and educational society dedicated to providing total water solutions ensuring the effective management of water.

In addition to sector-specific risk management standards that address cybersecurity and a cybersecurity assessment tool, AWWA provides reports and other guidance to support the water and wastewater sector's management of cybersecurity.

International Society of Automation—ISA (https://isa.org)

ISA's ISA99 standards committee develops and maintains the ISA/IEC62443 series of standards, the only international standard for security of industrial automation and control systems. ISA also offers training courses and a certificate program for industrial cybersecurity, as well as read-only access to the ISA/IEC62443 series of standards.

ISA Global Cybersecurity Alliance—ISAGCA (https://isagca.org/)

A collaborative forum to advance cybersecurity awareness, education, readiness, and knowledge sharing. Membership is open to any organization involved in industrial cybersecurity—end users, automation providers, system integrators, consultants, and government agencies.

ISA Security Compliance Institute—ISCI (https://www.isasecure.org)

The ISA Security Compliance Institute functions as an operational group within ISA's Automation Standards Compliance Institute. The Institute provides market awareness, technical support, education and, compliance for the ISASecure IACS security requirements that are based on ISA/IEC62443.

Cybersecurity & Infrastructure Security Agency—CISA (https://www.cisa.gov)

The CISA is the operational lead for federal cybersecurity and the national coordinator for critical infrastructure security and resilience. We are designed for collaboration and partnership. Learn about our layered mission to reduce risk to the nation's cyber and physical infrastructure.

The CISA website provides various resources including the National Vulnerability Database, the CSET vulnerability assessment tool, and other reports and guidance.

Infracritical (http://infracritical.com/)

Infracritical is an organization founded by Bob Radvonsky, Jake Brodksy, Tammy Olk, and Michael Smith, internationally recognized experts in the field of industrial cybersecurity.

Infracritical provides two resources:

- The Systems and Cyber Impact Database Markup (SCIDMARK) (directly available at http://search.infracritical.com/). This site provides a register of industrial cybersecurity incidents.
- The SCADASec mailing list, free to join, share knowledge and discuss industrial cybersecurity issues with experts around the world.

Top 20 secure PLC coding practices (https://www.plc-security.com/)

A project to provide guidelines to engineers that are creating software used in industrial control systems.

ICS Advisory Project (https://www.icsadvisoryproject.com/)

The ICS Advisory Project provides smaller operational technology asset owners, CISOs, cybersecurity analysts, and researchers with an analysis tool that allows them to quickly identify threats and vulnerabilities by product, vendor, and by critical infrastructure sector. Interactive dashboards collate publicly available threat and vulnerability data such as CISA ICS Advisories, CVEs, and MITRE ATT&CK.

Mitre (https://www.mitre.org/focus-areas/cybersecurity)

The MITRE Corporation was founded in 1958 as a not-for-profit company to serve as advisers in systems engineering to government agencies, both military and civilian. They maintain a database of vulnerabilities called CVE, the Common Vulnerabilities and Exposures database, and also provide cybersecurity guidance and frameworks.

Dale Peterson (https://dale-peterson.com)

Dale Peterson is the founder of security consulting firm Digital Bond and the S4 conference. He is an industrial control systems cybersecurity evangelist, and his blog posts and podcasts provide an excellent resource for those in the profession.

Ralph Langner (https://www.langner.com/)

Ralph Langner is recognized for his comprehensive analysis of the Stuxnet malware and is considered a leading expert in industrial control systems cybersecurity. His company's website provides a wide variety of resources for those in the profession.

(CS)2AI (https://www.cs2ai.org/)

Founded by Derek Harp and Bengt Gregory-Brown, (CS)2AI is a global, not-for-profit, workforce-development organization supporting professionals of all levels charged with securing control systems. They provide a platform for members to help members foster meaningful peer-to-peer exchange, continue professional education, and directly support cybersecurity professional development.

Dragos (https://www.dragos.com/)

Founded by Rob Lee, a former U.S. Air Force Cyber Warfare Operations Officer and renowned industrial cybersecurity expert, Dragos regularly provides detailed analysis of high-profile incidents such as TRISIS. The Dragos website is an excellent resource for those in the profession.

Dragos provides an Operational Technology—Cyber Emergency Readiness Team dedicated to addressing the operational technology resource gap that exists in industrial infrastructure. Designed to support asset owners and operators of industrial infrastructure, Dragos OT-CERT provides free cybersecurity resources for the Industrial Control System (ICS)/OT community (https://www.dragos.com/community/ot-cert/)

InfraGard (https://www.infragard.org/)

InfraGard is a partnership between the FBI and members of the private sector for the protection of U.S. Critical Infrastructure. InfraGard connects owners and operators within critical infrastructure to the FBI, to provide education, information sharing, networking, and workshops on emerging technologies and threats.

InfraGard's membership includes business executives, entrepreneurs, lawyers, security personnel, military and government officials, information technology professionals, academia, and state and local law enforcement.

ICS Advisory Project (https://www.icsadvisoryproject.com/

The ICS Advisory Project is an open-source project to provide DHS CISA ICS Advisories data visualized as a Dashboard and in Comma Separated Value format to support vulnerability analysis for the OT/ICS community. The ICS Advisory Project was created by Dan Ricci to provide smaller operational technology asset owners, CISOs, cybersecurity analysts, and researchers with an analysis tool that allows them to quickly identify threats and vulnerabilities by product, vendor, and even by Critical Infrastructure sector.

4.0 FURTHER READING

The following books are either referenced in this book or are recommended for those who would like to learn more about cybersecurity, how to assess and quantify risk, how to incorporate cybersecurity into operational activities in a WWU, how to help create a culture in which everyone understands cybersecurity risk, and the importance of physical security in an effective cybersecurity management system.

Barker, J. (2020). *Confident cyber security*. Kogan Page Ltd.

F.C. (2023). *How I rob banks (and other such places)*. John Wiley & Sons.

Hubbard, D. W., & Seiersen, R. (2016). *How to measure anything in cybersecurity risk*. John Wiley & Sons.

Langridge, M. (2022). *Information technology for water and wastewater utilities* (2nd ed.; Manual of Practice No. 33). Water Environment Federation.

Marszal, E., & Jim McGlone, J. (2019). *Security PHA review for consequence-based cybersecurity*. International Society of Automation.

Mustard, S. (2018). *Mission critical operations primer*. International Society of Automation.

Mustard, S. (2022). *Industrial cybersecurity: Case studies and best practices*. International Society of Automation.

Page-Bottorff, T. (2021). *Safety, health, and security standards for water resource recovery* (7th ed.; Manual of Practice No. 1). Water Environment Federation.

Radvanovsky, R., & McDougal, A. (2024). *Critical infrastructure, homeland security and emergency preparedness* (5th ed.). CRC Press.

Index

www.ingramcontent.com/pod-product-compliance
Lightning Source LLC
LaVergne TN
LVHW060558240526
839643LV00003B/42
* 9 7 8 1 5 7 2 7 8 4 7 1 0 *